ABOUT THE AUTHOR

IAN DEARDEN read philosophy at Downing College, Cambridge,
and Bedford College, London. He taught philosophy at Bedford
College, the City University, the Polytechnic of North London and
the University of Essex, and also for the London University
Department of Extra-Mural Studies.

Do Philosophers Talk Nonsense?

An Inquiry into the
Possibility of
Illusions Of Meaning

DO PHILOSOPHERS TALK NONSENSE?

AN INQUIRY INTO THE POSSIBILITY OF ILLUSIONS OF MEANING

IAN DEARDEN

DO PHILOSPHERS TALK NONSENSE?

First Published 2005

ISBN 1-904408-10-9

Published by TELLER PRESS
a division of Bank House Books

Printed and bound by Lightning Source.

Designed and typeset in England by
BANK HOUSE BOOKS
PO BOX 3
NEW ROMNEY
TN29 9WJ

Cover illustration by Pete Randle

CONTENTS

Chapter One: A New Kind Of Error.

For the best part of a century now philosophers have been accusing each other of talking nonsense. When one looks at what they mean by this one finds that they are assuming the possibility of a remarkable error, that of thinking there is something one means when there is not. I believe that this is an interesting and disputable assumption that has not received the attention it deserves. I admit that the view that philosophical problems (or some of them) and their proposed solutions (or some of them) are nonsensical is an ingenious response to their continued intractability but think it far from obvious that it is justified.

The view that it is possible to be mistaken in thinking one means anything by what one says does not appear to have a name, a state of affairs that is in itself remarkable. I shall call it 'nonsensicalism'. I shall refer to the alleged errors in question as 'illusions of meaning' (IOMs).

It is not clear whether the non-philosopher recognises the possibility of IOMs. Colloquially, the word 'nonsense' is often used to refer to the patently false: the accusation of talking nonsense then amounts to that of uttering outrageous falsehoods and this assumes that the accused is speaking meaningfully. No doubt it also happens that someone dismisses an impenetrable text, a piece of gobbledygook, as nonsense. But there need be no suggestion that its author is wrong to think he means anything. What is implied is that the speaker can make nothing of the text and is not going to waste further time on it. Consider the fact that one is constantly bombarded with defective utterances - ungrammaticalities, malapropisms, spelling errors, unfinished sentences, mumblings, slips of the tongue and pen, misprints. Often what is said or written is strictly nonsense, in that it is not part of the language. Yet in daily life it rarely crosses one's mind that the utterer does not mean anything, still less that he thinks there is something he means when there is not. One case where the non-philosopher *perhaps* allows for the possibility of IOMs concerns the utterances of the insane.

It is worth comparing the idea that people can be wrong in thinking they mean anything with the idea that they can have mental states of which they are unaware. Both are, if correct, extremely important. Both lead to 'Where will it all end?' worries in that one wonders how much of what one does is unconsciously motivated and how much of what one says is nonsensical in the relevant sense. Both lead to accusations against which it is difficult to defend oneself: if one is accused of harbouring an unconscious wish it is no use simply protesting one does not, and if one is accused of being mistaken to think one means anything by an utterance, it is no use just protesting that one does mean something. Finally, there may be significance in the fact that both ideas came to prominence at about the same time: certainly they are both radically anti-Cartesian, involving the claim that a person is not necessarily the best authority on his own mental life and that someone else might be in a better position to understand him than he is himself. [1]

But there is a striking difference. The notion of the Unconscious was from the first recognised as an innovation. People realised that they were being asked to accept something new and wondered whether they ought to: some refused. Freud, not of course the originator of the notion but surely the one who did most to place it on the intellectual map, was well aware of this and put much effort into arguing that the concept of unconscious mental entities was neither a contradiction in terms nor a mere *façon de parler* but one without which one cannot give a satisfactory account of the mental. I am not saying that the debate about the status of the Unconscious has always been of great intellectual rigour, but a debate has at least taken place. By contrast the notion of talking nonsense used by philosophers has provoked no similar discussion. The accusations made by the logical positivists did cause some annoyance but this seems to have been occasioned by what they chose to attack[2] and the

---

1. Though in the case of nonsensicalism it will be a peculiar kind of understanding, not a matter of understanding *what he says* but of understanding *his saying it.* The later Wittgenstein writes in a number of places that he understands how philosophers come to talk nonsense, that he is familiar with the moods and states of puzzlement from which such talk arises. P.M.S. Hacker (*Wittgenstein's Place in Twentieth Century Analytic Philosophy,* Blackwell, 1996, p.173) credits him with an interest in the pathology of philosophy - how philosophers are misled - which is lacking in, for example, Austin. It is important not to ignore this empathic aspect to Wittgenstein's diagnoses but we will have to ask whether it helps much with explaining how IOMs could occur and whether it is indeed possible to understand someone's saying something without understanding *what* he says.

2. One must not be swayed in one's assessment of nonsensicalism by one's attitude to the sorts of things nonsensicalists have attacked. Consider the positivist attack on religious utterances. Now, if nonsensicalism is tenable, it is surely *possible* that religious utterances (or some of them) are nonsensical. It is also possible that some attacks on religion are nonsensical. Indeed nonsense might occur anywhere in

criterion of meaningfulness they employed, not by the suggestion that people could be mistaken in thinking they meant anything.

If one considers the philosophers who have most often made accusations of talking nonsense (Wittgenstein and his followers, the positivists, Ryle and some other 'ordinary language' philosophers), it is hard to think of a single passage in which any of them expresses doubt about the correctness of nonsensicalism. I do not mean that they have done nothing to explain how IOMs might be possible - what has been done will be considered in due course - but I do find their supreme confidence astonishing. Philosophers who are quick to spot anything odd in the assertions of others see nothing odd in an assumption that underlies their own approach. Why, for example, does Wittgenstein in the *Investigations* never make the imaginary interlocutor, who voices objections to his points, ask, 'But is it *possible* to talk nonsense in the sense you think philosophers so often do?'?[3] Moore, in a well-known passage, said that it was not reflexion on the world that had provided him with philosophical problems but the assertions of other philosophers.[4] This book is concerned with a third thing that might cause puzzlement: what philosophers say about other philosophers.

Accusations of talking nonsense are not perhaps as common now as was once the case - in the heyday of logical positivism and again when the influence of the later Wittgenstein was at its height. But one does not have to look far to find philosophers, mainly Wittgensteinians, who are still prepared to make them. And this in itself prompts a question: Why do some philosophers do so, whereas others do not? What grounds, if any, do either have?

The philosophies of the *early* Wittgenstein and of the positivists are now widely rejected and there is a measure of agreement about what was wrong with them. It is not surprising therefore that

philosophy or outside it. (Wittgensteinians sometimes write as though only philosophers talk nonsense but this would surely not follow simply from the possibility of IOMs.) On the other hand, if nonsensicalism is mistaken, then the charge of talking nonsense is something that neither religious believers nor anyone else need worry about.

3. Wittgenstein once said that he regarded the *Tractatus* as the *only* alternative to his later work (*Ludwig Wittgenstein - A Memoir*, Norman Malcolm, O.U.P., 1958, p.69). Apparently then the only two philosophies with any hope of success are nonsensicalist ones.

4. 'An Autobiography', p.14, in *The Philosophy of G.E. Moore*, ed. P.A. Schilpp, 1942.

nonsensicalist accusations on grounds furnished by them are no longer often made. (This is no more than a broad generalisation: we shall see that verificationism, for example, is not completely dead.) But might there be other, more acceptable, grounds on which to make them? Might the philosophy of the *later* Wittgenstein or considerations as yet undreamed-of furnish such grounds? What puzzles me is that there has been, to my knowledge, no attempt to assess the general practice of making these accusations.[5] In so far as there has been any discussion of the matter the focus has been on the criterion of meaningfulness employed. But have these accusations *any* place in philosophy? Is it *possible* to be wrong in thinking one means anything? Might the defects of the *Tractatus* and of logical positivism be symptomatic of more general difficulties with nonsensicalism? Nonsensicalism seems neither to have been systematically defended nor to have been systematically attacked.

It may be that some philosophers think that IOMs are possible but that this does not have the major implications for philosophy that it was once thought to have.[6] Colin McGinn, for example, in *Problems in Philosophy: The Limits of Inquiry,*[7] clearly does not think that philosophical problems are *in general* likely to prove nonsensical: his thesis is that many, perhaps most, of them are genuine but not soluble *by us*. He is prepared however to make dismissive remarks that suggest that sometimes philosophers are duped by the meaningless. For example, concerning what be calls 'M-positions' (involving a supernatural metaphysics)[8] he writes:

> I take M-positions seriously, not as genuine candidates for truth, but as expressive of the philosophical hysteria that so readily

---

5. Many of the contributors to a recent anthology, *The New Wittgenstein* (ed. Alice Crary and Rupert Read, Routledge, 2000), come frustratingly close to asking the sorts of question I think need to be asked. I ponder this situation in Chapter Four.

6. One thing that suggests to me that few philosophers have consciously abandoned nonsensicalism is that when a philosopher is accused of talking nonsense it always seems to be assumed that the onus is on him to show that he does mean something (to 'give sense' to what he says) rather than on his accuser to explain how he could be mistaken in thinking he meant anything. Try the following experiment. Accuse some-one of talking nonsense because what he says is X (unverifiable, self-referential, with no use outside philosophy ... whatever seems best to fit the case). You will find, I think, that he will argue either that it is not X or that X-ness does not entail nonsensi-cality. He will not go on the offensive and demand to know how you think he could be wrong to think he meant anything.

7. Blackwell, 1993.

8. Ibid., p.16.

12

envelops us. What is interesting is that we can find ourselves uttering these words, or falling inchoately into these thoughts. For it is doubtful that any of this really means anything. It is mere poetry, rhetoric, word-spinning. Falsity is not the main problem, though doubtless there is some of that; the problem is rather that of coherence, of staking out a genuine position in logical space. What could it *mean* to say that consciousness is supernatural? What content does the notion of the supernatural really have?[9]

It would seem that he thinks that talking nonsense in the sense with which we are concerned is a possible philosophical error but just one among others; he would not endorse such famous programmatic remarks as that whenever someone says something metaphysical we must 'demonstrate to him that he had failed to give a meaning to certain signs in his propositions' (*TLP* 6.53)[10] or that '[t]he results of philosophy are the uncovering of one or another piece of plain nonsense and of bumps that the understanding has got by running its head up against the limits of language' (*PI* 119). My own feeling is that until we have asked whether the error is possible at all, we cannot have much confidence in *any* judgment about how widespread it is.

My task then is to persuade many philosophers that nonsensicalism needs to be questioned but perhaps also to persuade others that the issue is worth reviving. Since the book is concerned with the difficulties facing nonsensicalism, it could be said that it is primarily addressed to the first group. So to them I shall content myself for now with making one hortatory suggestion: surely, if one could successfully explain how IOMs are possible, this might have the added benefit of helping one decide *which* nonsensicalist accusations are justified. I don't suppose there is anyone, however sympathetic to nonsensicalism, who thinks that all the philosophical accusations of talking nonsense that have ever been made have been justified. One needs a way of assessing them and one question one might ask is, 'What explanation, if any, does the accuser have of how the accused was deceived and how good is it?'

I need perhaps to say more to the second group, those who may suspect me of trying to breathe life into a dead or dying issue.

---

9. Ibid., p.34.
10. Consider also in this connexion 4.003: 'Most of the propositions and questions to be found in philosophical works are not false but nonsensical ...[They] arise from our failure to understand the logic of our language.'

DO PHILOSOPHERS TALK NONSENSE?

However thoroughgoing one might believe the defeat of the *Tractatus* philosophy and of verificationism to have been (and there will be differences of opinion here), it clearly does not follow that accusations of talking nonsense should never be made. Perhaps Wittgenstein himself demonstrated what was wrong with his earlier philosophy but he did not think that he had shown that nonsensicalism was mistaken nor of course do contemporary Wittgensteinians think that he or anyone else has shown this. And even if the extensive literature on the failings of the verification principle is sufficient to show that no version of it is likely to be viable [11], nothing follows about whether there might be other grounds on which philosophers can more justly be accused of talking nonsense. I shall be arguing that the difficulties confronting verificationism have much to teach us about the difficulties that are going to confront other forms of nonsensicalism, but it will be a matter for investigation whether these latter can be overcome. Yet if any assumption of guilt by association must be avoided, so too must any assumption of innocence by dissociation. In particular the nonsensicalism of the later Wittgenstein cannot be assumed to be defensible just because it differs from earlier forms.

If there are philosophers who think nonsensicalism pretty well defunct, I suspect that there may be others who are not particularly familiar with it. Someone who began his study of the subject in the past two or three decades, even if his training was broadly in the analytical tradition, may have only a nodding acquaintance with the fact that for much of the Twentieth Century a revolution was widely felt to be under way, in which the new techniques of logical and linguistic analysis were being put to work to dissolve the hitherto intractable problems of philosophy once and for all. To an earlier

---

11. Perhaps the most frequently discussed difficulty is that of formulating the principle so as to exclude metaphysics without also excluding scientific laws. Whilst not considering this problem to be totally without interest, I think that excessive concentration upon it has done as much as anything to divert attention from the question whether the error the principle is supposed to detect is possible. Even if there were no difficulty in demarcating the unverifiable, it would still not be obvious that one should pass from 'unverifiable' to 'meaningless' nor how one could think one meant something by a meaningless (because unverifiable) claim.

12. Gollancz, revised edition, 1946, Chapter VIII. It is clear that Ayer is not offering *solutions* to genuine problems but *dissolutions* of pseudo-problems. The self-confidence of the chapter's title is surpassed on p.134 where he describes the questions as having 'played a part in the history of philosophy which is out of all proportion to their difficulty or their importance.'

generation the seminal importance of Russell's Theory of Types, the impact of the *Tractatus*, the unsubtle response of the positivists and the sophistications of Wittgenstein's later philosophy are a familiar story; and the fact that Ayer could entitle a chapter of *Language, Truth and Logic* 'Solutions of Outstanding Philosophical Disputes'[12] might now seem like *hubris* but it will be a familiar *hubris*, understandable in the context of its time. But more recently trained philosophers might have some difficulty entering into the nonsensicalist way of thinking, at least when it is presented as appropriate to *all* philosophical problems. I have not in this book attempted a historical account of the development of nonsensicalism[13] but I hope I have given a sufficient number of examples of the nonsensicalist approach for the reader to get some feel for it.

There are in fact several reasons why the issue of nonsensicalism is worth discussing that ought to appeal to philosophers of all persuasions. First, it is of importance for both the philosophy of language and the philosophy of mind to determine whether IOMs are possible.[14] It is especially important for the area of overlap of these two branches of the subject: the investigation of the relationship between thought and language. Second, are not nonsensicalists providing, a forthright and challenging answer to the question, 'Why does language matter to philosophy?'?[15] Third, in Chapter Six I shall argue that the issue is of more than narrowly philosophical interest. Fourth, it is wrong to think of nonsensicalism as like a weapon which one may or may not choose to use. If a putative claim really is

---

13. A historical discussion of nonsensicalism would have to take account of the fact that nonsensicalist-sounding remarks can be found in several pre-Twentieth Century philosophers, in Aristotle, Hobbes, Locke, Berkeley, Hume, Kant, and no doubt others. In the empiricists particularly one can even discern rudimentary theories about how philosophers can think they see sense in nonsense. Nevertheless, I am inclined to think that it was only in the Twentieth Century that language came to be seen as sufficiently central to philosophy to make plausible large-scale dismissals of philosophical questions and theses as nonsensical.

14. Some philosophers would no doubt approach the problem by producing their own account of what meaning is and deducing from that whether IOMs are possible. It seems to me that too little about meaning is obvious or uncontroversial for this approach to carry conviction. I believe that investigating the question of IOMs is something to be done first if we want a soundly based account of meaning. However, in Chapters Eight and Nine I do consider the possibility of using a Wittgensteinian account of meaning (more precisely, of what meaning is *not*) as an explanation of how IOMs might be possible.

15. See Ian Hacking's book of that title. C.U.P., 1975.

nonsensical, then certain criticisms of it - those that treat it as meaningful but false - will be just as misguided as the claim itself. (In Chapter Three I shall suggest that if an utterance is nonsensical, this is the *only* criticism that can be made of it.) So no philosopher can afford to neglect the question whether he should be prepared to encounter IOMs.

A brief guide to the structure of the book may be helpful, the more so as I leave the most obvious question – how, if at all, are IOMs possible? - until the later chapters.

Chapter Two uses Malcolm's book *Dreaming* to illustrate certain difficulties in referring to nonsense and accusing others of talking nonsense. One reason I choose this work is that almost everyone takes the view of dreaming that Malcolm claims is nonsensical. Almost everyone is thus on the receiving end of his nonsensicalist accusations.

Chapter Three generalises the findings of Chapter Two, arguing that nonsensicalists must submit to an obvious but severe constraint: they must not treat nonsense as if it were meaningful and merely an especially horrendous falsehood. This creates difficulties, for example in specifying just what it is that is being rejected as nonsense and in showing that others are in fact victims of IOMs.

Chapter Four asks whether such difficulties arise because I impose an over-strict conception of nonsense on the nonsensicalists. I argue that if there is to be anything distinctive about their approach, they are committed to the possibility of IOMs and I spell out what this involves.

Chapter Five discusses the 'staying power' of alleged philosophical nonsense: the utterances in question do not normally lose their *appearance* of meaningfulness to those who officially condemn them as nonsense. Might this be because they are not nonsensical at all? The problem is closely related to that of the non-existence of obvious and uncontroversial examples of philosophical nonsense.

Chapter Six asks whether IOMs, if they are possible, might occur outside philosophy - do dreamers, schizophrenics, the takers of certain drugs and the victims of certain hoaxes suffer from them?

Chapter Seven asks how IOMs might be possible. It begins by stressing how odd they must be: they must be illusions without content in that one cannot say *what* the victim of one thinks he sees. However it is not clear that this makes them impossible. Various suggestions as to how they might be possible are then reviewed. Most come from the later Wittgenstein.

A NEW KIND OF ERROR

None of the ideas discussed in Chapter Seven ever seems to have been fully worked out. Why this seemingly astonishing perfunctoriness? In Chapter Eight I suggest that at least for Wittgenstein the answer might be that he believed that when a person means something by an utterance this is not a matter of what is going on in his mind at the time and that therefore in principle *anything* that goes on in a genuine case of meaningful speech could also go on in an IOM.

Chapter Nine argues that if this was his view, everything will then depend on the reasons given in a particular case for denying that someone means anything by an utterance. I consider general criteria of meaningfulness and more piecemeal approaches but find them inadequate: for example, they often involve linguistic legislation, arbitrary stipulation of what is to count as meaningful. I also consider the possibility that the accused might come to admit that he has been talking nonsense.

Chapter Ten attempts to reach a general conclusion about the defensibility of nonsensicalism.

I begin my discussion with a consideration of Norman Malcolm's book *Dreaming*.[1] I choose this work for two reasons. First, Malcolm writes clearly and a particular habit of nonsensicalists is strikingly obvious in his case: the tendency to discuss alleged nonsense as if it made perfect sense but were defective in some other way. Second, almost everyone finds Malcolm's thesis incredible, not excluding philosophers who, like him, are influenced by Wittgenstein.[2] Since he regards the natural, naive, pre-Malcolmian view of dreaming as *nonsense*, then pretty well all of us, because we hold that view or at least think we understand it, would be accused by him of talking or being deceived by nonsense. His argument should bring home to philosophers how mystifying it can be to be accused of talking nonsense, even to those who are not averse to making such accusations themselves. If I had chosen another issue where such accusations abound - theism, sense-data, private language, the mind-body relation - this would not necessarily have been so.

Most of what I say explicitly will concern the first reason. Generally speaking, I shall allow the paradoxical quality of Malcolm's accusations to make itself felt without help from me and only draw attention to it when there is some special reason for doing so.

Let me summarise Malcolm's main theses:

1. It is nonsensical to assert that one is asleep since the fact of one's making assertions shows that one is not asleep. For similar reasons, one cannot *judge* that one is asleep (or indeed make any judgment whatsoever) when one is asleep. Any attempt to give sense to the idea founders on the impossibility of verifying both that the mental activity is occurring and that the person is asleep. As he puts

---

1. Routledge and Kegan Paul, 1959.

2. See, for example, D.F.Pears, 'Professor Malcolm: Dreaming', *Mind*, LXX, 1961, and J.F.M. Hunter, 'Some Questions About Dreaming' in *Essays After Wittgenstein*, University of Toronto Press, 1973. Malcolm's treatment of dream-scepticism has perhaps been more favourably received than his treatment of dreaming as a whole. (See H-J. Glock, *A Wittgenstein Dictionary*, Blackwell, 1996, pp.340-1.) A view of dreaming that seems quite close to Malcolm's and which also derives from Wittgenstein's later writings is put forward by Graham McFee ('The Surface Grammar of Dreaming', *Proceedings of the Aristotelian Society*, 1994), though he avoids the use of 'nonsense' and similar words. If this avoidance is deliberate and MacFee is not a nonsensicalist, the resemblance to Malcolm may be only superficial.

it, there is *no criterion* for mental activity during sleep.

2. To the claim that we *dream* during sleep Malcolm replies that we use the word 'dream' as follows: someone awakes seeming to remember events that never occurred and we say that he *dreamed* them. That—the waking report—is the criterion of having dreamed. Only in this sense do we allocate dreams to particular periods of sleep and it does not follow that the dream had a more precise location and duration in physical time. The verificationist considerations in (1) show that it is nonsense to speak of the dream as occupying a definite substretch of the time spent sleeping. They also show that there is no sense in the idea that one could awaken with a faulty memory of what one dreamed.

3. 'I am dreaming' is as nonsensical as 'I am asleep' since it would imply 'I am asleep'. There is therefore no problem of dream-scepticism since the suggestion that one might be dreaming (and that one therefore needs a way of telling whether one is awake or dreaming) is nonsensical.

It is only upon one aspect of Malcolm's arguments, his use of the notion of nonsense, that I shall focus and I shall not pursue certain problems stressed by other critics, e.g. the implications of the fact that he seems to limit many of his claims to 'deep' sleep and the question whether his attempt to pin down the concept of dreaming by a *single* criterion is necessitated by—or even faithful to—Wittgenstein's use of the term 'criterion'. I shall however discuss Malcolm's verificationist use of that term.

Malcolm's view of dreaming has been thought a *reductio ad absurdum* of the methodology of the later Wittgenstein. Hacker rightly sees this as too hasty.[3] Nevertheless, I suggest that the difficulties that Malcolm's nonsensicalism faces ought to make us at least wary of other forms, including Wittgenstein's.

It might well be wondered what Wittgenstein would have thought of Malcolm's view. Perhaps he would have said that the 'surface grammar' of dreaming strongly suggests that it is an experience during sleep, occurring at a definite time, involving the dreamer's having thoughts and remembered in the ordinary sense of 'remember', but added that 'surface grammar' can mislead. To this extent he might have supported Malcolm. But unless he agreed with Malcolm's single-criterion approach and his verificationism, he need

---

3. P.M.S. Hacker, *Wittgenstein's Place In Twentieth-Century Analytic Philosophy*, Blackwell, 1996, p.239.

not have said that the view insinuated by 'surface grammar' could not be right. Perhaps a consistently Wittgensteinian view would be agnostic here: we use the word 'dream' as we do and this does not depend upon whether the view that 'surface grammar' naturally inclines us to take is correct.[4]

Many philosophers would, I think, say that it is possible though perhaps unlikely that empirical discoveries could undermine the ordinary view of dreaming. But this only emphasises their disagreement with Malcolm: the ordinary view, far from being conceived of as nonsensical, is regarded as probably but not certainly true.

Malcolm's Verificationism.

Malcolm relies heavily on verificationist arguments:

> I have stressed the senselessness, in the sense of impossibility of verification, of the notion of a dream as an occurrence 'in its own right', logically independent of the waking impression, and to which the latter may or may not 'correspond'.[5]

Sometimes the verificationist demand is applied not directly to a claim but to the question whether someone understands that claim:

> It is a logical impossibility that there should be a criterion for saying that someone understands how to use the sentence 'I am asleep' to describe his present state. This is equivalent to saying that the idea of such a use is not intelligible.[6]

This passage also shows that Malcolm sometimes formulates his verificationist arguments using the term 'criterion'.[7] Why is he so confident that the claims he attacks are unverifiable?

---

4. For an entirely different view of Wittgenstein on dreaming see McFee, op.cit.

5. Op. cit., p.83.

6. Ibid., pp.16-17.

7. See ibid., p.109 for another passage illustrating both these points. There is a slight possibility of confusion with the word 'criterion'. In Chapter One, I spoke of the logical positivists' use of verifiability as a criterion of meaningfulness. I am now speaking of Malcolm's verificationist use of Wittgenstein's notion of criterion (of which more later in this chapter). Thus one could, I suppose, say that for Malcolm it is a criterion (in the ordinary sense) of the meaningfulness of certain utterances involving the word 'dream' that there should be a criterion (in Wittgenstein's semi-technical sense) for their truth. In practice I do not think that anyone familiar with the subject matter would be confused, but it is worth drawing attention to the possibility.

We noticed that it would be self-contradictory to *verify* that a man was both asleep and judging that he was, because whatever in his behaviour showed that he was making the judgment would equally show he was not asleep. Now this would be so *whatever* the judgment was. In order to know that he had made any judgment one would have to know that he had said certain words and that he had been aware of saying them. But whatever it was in his demeanour that revealed his awareness of saying them would also establish that he was not asleep.[8]

However his conclusion concerns nonsensicality, not self-contradiction:

It could be objected that my argument has shown merely that the *verification* that someone is both asleep and judging is self-contradictory, not that his *being* both asleep and judging is self-contradictory. This is true. The latter notion is not self-contradictory (in the sense of entailing both of a pair of contradictory propositions). But it is senseless in the sense that nothing can count in favour of either its truth or its falsity.[9]

How reliant is Malcolm on verificationism? An argument that seems not to be verificationist appears on pp.6-7. One could no more assert that one was asleep than one could assert that one was dead. One could say the words 'I am asleep' or 'I am dead' but one would not be making assertions because one can only assert what could conceivably be true (a questionable claim but I shall let it pass). Now Malcolm realises that someone might challenge the analogy between being asleep and being dead: perhaps one can be in a state of consciousness when asleep (even if one is not conscious of events in one's vicinity) and perhaps one could then make assertions or at least judgments. And to rebut this he falls back on verificationism:

Neither I nor anyone else can *find out* whether the state of myself that I claim to describe by the sentence 'I am asleep'

---

8. Ibid., p.36.

9. Ibid., pp.36-7. In his earlier article 'Dreaming and Scepticism', *Philosophical Review*, LXV, 1956, Malcolm seems much more inclined to assimilate 'I am asleep' to contradictions, though he stops short of calling it simply an ordinary contradiction. By the time of *Dreaming* this tendency survives only in his view that it is contradictory to speak of *verifying* that someone is both asleep and judging, and in his curious claim that '[i]f a man could assert that he is asleep, his assertion would involve a kind of self-contradiction' (p.7).

really is the state of being asleep.[10]

To sum up, Malcolm is quite seriously rejecting as nonsense the idea of mental activity during sleep and his reasons are avowedly verificationist:

> That someone reasoned, concluded, believed, tested, pondered, perceived, knew, decided, something while asleep - would all be assertions without meaning in the sense that nothing could count for or against their truth.[11]

The Natural Response to Malcolm's Verificationism.

To someone unconvinced by Malcolm's approach it will seem that he has pointed out real difficulties in ascertaining that someone engaged in mental activity during sleep—but that is all. His critics normally discuss such things as whether talking in one's sleep and the fact that events in the sleeper's environment sometimes seem to be 'woven' into his dreams count as evidence for such activity. But I want rather to ask: no matter how great the difficulties in *verifying* that someone is judging, say, during sleep, ought one to conclude that if the suggestion is unverifiable, it is nonsensical? Does it not look as though Malcolm must understand what possibility is under discussion in order to argue that the realisation of that possibility is unverifiable? He argues that no one could be in a position even to make *past tense* judgments that he made an assertion or judgment during sleep, because this would have to be the result of an inference and

> ...we always run into the difficulty that we do not know what the goal of the inference is, because we are not in possession of any criterion for saying that a man made a judgment while asleep...[12]

One is inclined to retort that Malcolm knows perfectly well what the goal of the inference is and that is why he is able to marshal arguments to show how difficult it is to reach that goal. This would still be so, even if he were to produce *conclusive* arguments that it is *impossible* for anyone to have *any evidence whatsoever* that someone made a judgment while asleep.

The trouble is that if one considers an apparent possibility and then argues that the realisation of that apparent possibility would be

---

10. Ibid., pp.13, emphasis added.
11. Ibid., p.110.
12. Ibid., p.44.

unverifiable, it is not clear that one can *ever* conclude that talk of that possibility is nonsensical.[13] One has already accorded meaning to it and it seems too late to claim that nothing was meant. According sense to something in order to prove that it makes no sense smacks of the mistake Wittgenstein warns against at *PI* 500: saying of a senseless locution that its sense is senseless.[14]

Nonsense And Inference.

Malcolm persistently talks as though nonsense could stand in logical relations with other nonsense. For example:

> ... the famous philosophical question, 'How can I tell whether I am awake or dreaming?' turns out to be quite senseless since it implies that it is possible to *judge* that one is dreaming, and this judgment is as unintelligible as the judgment that one is asleep. Furthermore, the question appears to presuppose that one might be able to *tell* that one is dreaming, which is double nonsense: for this would mean that one made an inherently unintelligible judgment while *asleep*.[15]

> If 'I am dreaming' could express a judgment it would imply

---

13. Consider 'near-death experiences' - visions of a vaguely religious character recalled by those who have been close to death or even pronounced dead. It is sometimes alleged that they occurred at a time when brain-activity had ceased. This faces the Malcolmian problem, 'Might one not be dealing with false memory-impressions of experiences that never occurred?' and the (to my mind) more obvious problem, 'Perhaps they did occur but either before or after the period of no brain-activity.' The problems about verification are serious but it is far from obvious that they license the conclusion that we do not really know what possibility is being canvassed.

14. Malcolm in *Dreaming* does not, to my knowledge, make the mistake of explicitly speaking of a 'senseless sense' or 'meaningless meaning' but in 'Dreaming and Scepticism' he writes:

> ...it is not true, but *senseless*, to say that sound sleep and waking are 'indistinguishable' from one another, or that they are 'exact counterparts'. For the *meaning* of this philosophical remark is that identically the same sensations, impressions, and thoughts could occur to one in either condition (op. cit., p.25, emphases added).

15. *Dreaming*, pp.109-10. The passage continues, 'But those who hold to the coherence principle must bear an even heavier weight of nonsense, if that is possible'. (The extra nonsense is that if someone '*perceived* something and *tried* to connect it with the rest of his life ((as he *remembered* it)) but *saw* that it did not fit in, then he would be *assured* that he was *asleep!*') The phrase 'if that is possible' perhaps betrays a doubt whether nonsense can really pile up in the way suggested. A thesis can no doubt sometimes be unpacked into a *mélange* of implications and assumptions all of which are false, but what if anything corresponds to this when nonsense rather than falsehood is in question is not obvious.

23

the judgment 'I am asleep', and therefore the absurdity of the latter proves the absurdity of the former.[16]

Now it is clearly wrong to claim that something can be meaningless and yet imply or be implied by something else. But it is also wrong to say, as Malcolm appears to do in the second quotation, that *if* something made sense, it would imply such-and-such. What it implied would surely depend upon what sense it made. It does not yet mean anything (let us assume) and could presumably be given any sense whatsoever. Occasionally Malcolm seems to realise this:

> It ought not to be supposed that what has been shown is merely that there is something queer about the words 'I am asleep' which keeps them from expressing a judgment - and that it is *possible* to judge that oneself is asleep. For what is the description of this possible judgment? The judgment that would be expressed by the words 'I am asleep' *if* those words had sense? That is not the description of a judgment.[17]

What is Malcolm doing then? Perhaps he is assuming that 'X is dreaming' implies 'X is asleep' and trying to apply this principle to 'I am dreaming'. He is, as it were, making a formal 'inference', mechanically performing the substitution that seems licensed by the principle. He then looks to see whether the result makes sense and concludes that it does not do so. There is a danger here. If one arrives at nonsense by mechanically applying the principle, might this not be a reason for mistrusting the principle? But Malcolm has a possible defence. He might say that those he is attacking, adherents of the ordinary view of dreaming, also accept the principle and would 'reason' thus, and he is trying to use their way of thinking to get them to see they are talking nonsense. (I, for example, would accept the inference that if I am now dreaming, I am asleep.)[18]

But even if Malcolm could reformulate his arguments so as to avoid giving the impression that he thinks that nonsensical locutions

---

16. Ibid., p.109.

17. Ibid., p.35.

18. Two brief comments are in order. First, there is another respect in which Malcolm will have to be entering into the thought processes of those he thinks talk nonsense, for, given his view of dreaming he cannot really treat the present continuous 'X is dreaming' as analogous to 'X is walking', implying that a datable occurrence is in progress. Second, as we shall see in the next chapter, any nonsensicalist will have to provide some account of how it is possible for someone to think he is inferring '*p*' from '*q*', when '*p*' and '*q*' are nonsense.

can imply and be implied by other locutions, it is important to ask why he makes the slips he does. The obvious answer is that the alleged nonsense still *seems* meaningful to him, in spite of his official view, and still *seems* to him to have logical properties. Perhaps he would admit this but insist that the appearance is illusory. Philosophical nonsense would then be liable to seem meaningful and to tempt people into error even after they have unmasked it. But it is also possible that Malcolm really does understand the locutions in question and it is his view that they are nonsense that is mistaken.

Am I Being Fair To Malcolm?

Before proceeding further I want to consider certain objections to my treatment of Malcolm that might occur to the reader. Perhaps it will be felt that some of my criticisms are pedantic. It might be said, 'We surely know what Malcolm means, even if he does sometimes slip into talking of nonsense as if it were just a bad case of falsehood.' My reply is to repeat that it is not always clear how he could reformulate his points to take account of my criticisms and that his errors suggest that he himself seems to see sense in what he claims is nonsense.

Another worry the reader might have is that I am imposing on Malcolm too strict a conception of nonsense, one that is not his. This question I shall discuss in Chapter Four with regard to all those I dub 'nonsensicalists': what exactly do they imply by their accusations? Here I shall just make a few points about what Malcolm means.

First, he does distinguish the nonsensical from the false. His use of words like 'unintelligible' (p.44), 'not intelligible' (p.16) and 'without meaning' (p.110) to describe nonsense shows that he wishes to avoid the colloquial use of 'nonsense' to describe the manifestly false, as does a passage like this:

> My contribution (if it is one) to this renowned sceptical problem is to try to show that the sentence 'I am not awake' is strictly senseless and does not express a possibility that one can think.[19]

I suggested above that he has to accord meaning to the claims he is dismissing in order to apply his verificationist demand to them. Let us consider this further. He is trying to show that those

---

19. Ibid., p.118.

who say certain things are talking nonsense. But it is insufficient for him to mean by 'talking nonsense' either:

(a) that he does not understand them

(b) that these utterances are not acceptable sentences of the language.

Concerning (a), it is clear that there would be no philosophical achievement in claiming not to understand another philosopher (which is not to say that such claims are not often true).

How about (b)? It is sometimes possible to say with certainty that someone has uttered a locution that is not an acceptable sentence of the language. But philosophers' utterances are rarely like this. If 'I am asleep' or 'I do not know whether I am awake or dreaming' are bad English, this is not evident to someone investigating the matter equipped only with the grammar he learnt at school. I am not entirely clear what Malcolm's view is on their acceptability as sentences of English. On the one hand he writes:

> If I say 'He is sleepy' of someone, I make an assertion that entails the assertion he would make if he said 'I am sleepy'. There is not this relationship between 'He is asleep' and 'I am asleep'. If someone said the latter either he would be making no assertion at all or else he would be using his words in a different sense, e.g. to mean that he does not wish to be disturbed. 'I am asleep' does not have a use that is homogeneous with the normal use of 'He is asleep'.[20]

Here he seems to accept 'I am asleep' as a possible sentence and even suggests a use for it. Yet he also writes:

> The sentence 'I am asleep', no matter how respectable in appearance, was shown to be an inherently absurd form of words.[21]

But surely the exact status of 'I am asleep', considered as an English sentence, is of no great importance. Even if it could be proved to be bad English this would not show that those uttering it meant nothing by it, which is what Malcolm must show. If he admits that they mean *something*, then even if they have expressed themselves in a way that grammar and usage do not allow, attention

---

20. Ibid., p.6.

21. Ibid., p.109. No doubt the best way of reconciling the two passages is to assume that Malcolm thinks that it is the very absurdity of 'I am asleep' that suggests a possible use, i.e. it is a *jocular* way of saying that one does not wish to be disturbed, a grammatical joke no doubt. But this does not tell us whether Malcolm thinks that 'I am asleep' is a genuine English sentence.

will switch to *what they mean*. Once one has elicited an acceptable formulation, one can discuss that: whatever else might be said against it, it will not be nonsense.

Consider now Malcolm on 'I am dreaming':

> ...we know how to use the words 'I am awake' but not the words 'I am dreaming'. To speak more exactly, we know that ' I am dreaming' is the first person singular present indicative of the verb 'dream', and that dreaming and waking are logical contraries, and therefore that 'I am dreaming' and 'I am awake' are logical contraries. In this sense we know how to use the sentence 'I am dreaming'. On the other hand, considerations previously mentioned bring home to us that it can never be a *correct* use of language to say (even to oneself) 'I am dreaming'. In this sense we do not know how to use those words.[22]

I wonder if Malcolm really does think it could *never* be correct to say, 'I am dreaming'. He points out that we use 'Am I dreaming?' and 'I must be dreaming' as 'exclamations, expressing sharp surprise at some appearance of things'.[23] Surely there would be nothing odd in someone's using 'I am dreaming' similarly; it would merely be less idiomatic. If so, then not only can 'I am dreaming' be parsed, it might have a use. So again, what Malcolm is accusing others of doing is not so much uttering locutions that are not part of the language as suffering from IOMs when they utter certain locutions (which may or may not be part of the language). They say things like, 'I don't know whether I am awake or dreaming', mistakenly thinking that they themselves mean something by them. Sometimes what helps to mislead them is that the locutions do have a use and are not excluded from the language:

> It may be that part at least of the peculiar force of the philosophical question 'How can I tell whether I am dreaming now?' comes from our mixing up the actual use of the question, 'Am I dreaming?' with what, in our philosophical thinking, we imagine *ought* to be its use. As a result we confuse the sometimes sensible question 'How can I tell whether that thing over there is actually the way it looks to be?' with the always senseless question 'How can I tell

22. Ibid., p.114.
23. Ibid., pp. 20-1.

whether I am awake?'[24]

To sum up, I do not think I am being unfair to Malcolm by assuming that his accusations of talking nonsense involve his attributing IOMs to the accused and that he cannot consistently ascribe meaning to what someone says and call it nonsense or treat one piece of nonsense as entailing another.

Malcolm On The Coherence Principle.

Perhaps the most striking case of Malcolm's seeming to understand what he claims is nonsense - indeed of his being able to think in terms of it with as much ease as those who believe it makes sense - occurs in his discussion of the coherence principle. This is the traditional defence against dream-scepticism. Making use of it consists in

> ...noting whether certain 'phenomena' presented to one are connected in the right ways with other phenomena, past, present and future. The objection that should occur to anyone is that it is possible a person should *dream* that the right connexions hold, *dream* that he *connects* his present perceptions 'with the whole course of his life'.[25]

He goes on:

> But now I wish to make a criticism of it that is more consonant with the thought of this monograph.[26]

This second criticism of course involves his claim that it is *nonsense* to assert that one is or might be dreaming. But let us leave that and consider the first. Malcolm clearly thinks he is *accepting for the sake of argument* an assumption made by those worried about dream-scepticism and then showing on the basis of that that the coherence principle does not work. This would be unexceptionable if he were accepting for the sake of argument an assumption he thought *false*. We often argue, 'You are mistaken in believing that *p*, but even if it were true that *p*, it still would not follow that ...; you still would not have explained how ...; you would still face the objection

---

24. Ibid., pp.116-17. I do not think it wrong to hold that idioms can mislead. I myself believe that if the word 'nonsense' had never come to be used colloquially to refer to the manifestly false, it would have been far more obvious that philosophical accusations of talking nonsense are something radically innovative, deserving the closest scrutiny. But whether idioms can mislead one in Malcolm's sense - into thinking one means something when one means nothing - is another matter.

25. Ibid., p.108.

26. Ibid., p.109.

that ..., etc.' But Malcolm is criticising utterances he thinks are
nonsensical and it is doubtful whether he can simply adopt 'a
nonsensical assumption', whether he can say:

> The claim that I might now be dreaming is nonsense. But let
> us assume it makes sense. The coherence principle will then
> fail as a way of telling whether one is dreaming because ...

Malcolm apparently thinks he can enter empathically into the
minds of those who misguidedly see sense in nonsense. (He
expresses surprise that his objection has not occurred to proponents
of the coherence principle.[27]) Perhaps something can be done along
these lines; but he gives the impression that he is inquiring into what
would be the case if something that is nonsense (not: false) made
sense (not: were true) - as though he thinks there are
'countersensicals' as well as counterfactuals.

Malcolm's discussion of the coherence principle also shows
something important about how things seem to those accused of
talking nonsense. Those who (like myself) are unreconstructed
adherents of the pre-Malcolmian view of dreaming will be much
more impressed by his first argument than by the one relying on the
claim that it is nonsense to suggest that one is dreaming. Since *they*
believe it makes sense to talk of the possibility that one is dreaming,
they can take his first argument straight, without the above
misgivings. And indeed it does seem a strong one, one that ought to
give pause to anyone inclined to think that one can tell one is not
dreaming by noting that one's present experiences cohere with one's
past. But the point I want to make is that it seems ambiguous: there
are two ways of taking the claim that one might '*dream* that the right
connexions hold'. The suggestion might be that one could have a
dream that actually did cohere with one's past, without ceasing to be
a dream. Thus a man on Sunday night might dream of a typical
Monday morning, just such a Monday morning as might follow that
Sunday night. Or the suggestion might be that one could have a
dream that does not cohere with one's past though one does not
realise this at the time. Adults often dream that they are back at
school, without this striking them as anomalous at the time. Both
these interpretations seem to me to point to real difficulties with the
coherence principle.

Now I have deliberately expressed them in language to which

27. Ibid., p.108.

Malcolm would object (e.g. the phrase 'at the time'). I have done so because I want to make it clear how things seem to the non-Malcolmian. To him it makes sense to wonder whether one is dreaming. The coherence principle suggests itself as a plausible way of proving to oneself that one is not. Malcolm offers a perfectly intelligible argument that it does not work. On reflexion it emerges that there are two ways of taking what he says but both are strong arguments.

Malcolm must claim that all this is illusion. Not only is one bemused by a nonsensical pseudo-problem, one takes a nonsensical pseudo-principle to be an answer to it; that is, until another nonsensical consideration convinces one that one is mistaken. This nonsensical consideration on examination appears to resolve itself into two separate arguments, though both of these are in fact nonsensical. It is not obvious how all this is possible.

Attributing 'Nonsensical Assumptions' To Others.

We saw above that there are problems with the idea of adopting 'a nonsensical assumption' for the sake of argument. It is equally difficult to see how one can attribute 'nonsensical assumptions' to others. Someone can be guilty of assuming a falsehood, and the falsehood can be deeply buried, for it may take much argument to show that it is being assumed. But if one tries to do this with a 'nonsensical assumption', one will be up against the difficulty of treating nonsense as standing in logical relationships. Malcolm in fact accuses Freud of 'supposing a number of things which [should be] rejected as nonsensical[28], but inspection of the relevant passages in Freud shows that the assumptions in question are made fairly overtly, and so we perhaps only have a case of what Malcolm would consider to be *talking* (not *assuming* or *presupposing*) nonsense. I shall therefore not press the point against Malcolm; a far clearer case of the attribution of 'nonsensical assumptions' to others will be presented in the next chapter.

Is The Source Of Malcolm's Errors His Verificationism?

Even if the accusation of talking nonsense is a legitimate move in philosophy, it is clear that Malcolm has not fully adapted to the requirements that this form of polemic imposes. Perhaps he is

---

28. Ibid., p.121.

insufficiently rigorous and could with more careful wording of his arguments sidestep my criticisms. Or perhaps his errors are symptomatic of something deeply wrong with nonsensicalism. I shall say no more about this until I have generalised these problems to take in nonsensicalism as a whole. But it might be suggested that it is his *verificationism* that is at fault and I shall say a little more about that.

If there is one thing that can be learnt from the many rebuttals of Malcolm on dreaming, it is that one can disagree with him while still remaining close to Wittgenstein. Most writers have felt that Malcolm takes an unduly narrow view of what we accept as evidence for dreaming and many would add that it is un-Wittgensteinian to try to characterise dreaming by a single criterion and to neglect the uncertainty, stressed by Wittgenstein, about what is criterion and what is symptom.[29] I have no wish to slight the excellent work that has been done along these lines, but I wonder whether a more fundamental question has been neglected.[30]

Suppose there were no evidence for dreaming other than the waking report. Suppose sleepers never showed any signs of having experiences or of being influenced by events in their vicinity. One could not then accuse Malcolm of exaggerating the empirical inaccessibility of dreams, conceived of as experiences during sleep. But would he be justified in calling it *senseless* to suggest that dreaming was an experience during sleep, that people did not just awake seeming to remember things that never happened but awoke with genuine memories of experiences during sleep? It seems to me that one would still understand the suggestion and Malcolm would still have to understand it in order to argue for its unverifiability.

It is worth comparing Malcolm's verificationist dismissal of experiences during sleep with a more recent argument. P.M.S. Hacker writes:

> A.R.Luria in *The Mind of the Mnemonist* ... relates how the mnemonist explained an error in his mnemonic performance ... manifesting his ability to recollect a large number of random objects ... His mnemonic device was to imagine himself walking down a street in St. Petersburg, and as each object was called out, he would imagine himself placing it at a

29. See *BB* p.25; *PI* 79.
30. One writer who does raise it is Pears (op.cit.).

particular place in the imagined street ... [He] would recall the objects named by imagining himself walking down the street again, and would, as it were, read off the list of objects from his imagined scene. On one occasion he forgot an item, viz. a milk-bottle. He explained this by claiming that he had imagined putting the bottle ... in front of a white door, so that when he imagined walking down the street again ... he did not *notice* the milk-bottle against the white door! This makes *no sense*. (What would be the criterion for its *being there*, even though he did not 'notice' it?)[31]

Now I cannot think of much in the way of evidence for its being there. Perhaps neurophysiological correlations might help, although if 'anomalous monism' is anywhere near the truth, this may be a vain hope.[32] Or perhaps if the mnemonist regularly made mistakes of this sort - claiming to have overlooked red objects against red backgrounds and so on - this might be evidence. But it would hardly *establish* that the imagined bottle was there; I do not think Hacker would call it a 'criterion'. Does that make the explanation nonsense? I do not think so. Once again it seems one has to understand the claim in order to decide whether it can be verified.

Another writer who sympathises with the mnemonist is Brendan Wilson - 'I think his story *does* make sense (because, even if it is an excuse, it would be a poor one, if it were senseless).'[33] Hacker could reply that if the mnemonist can be taken in by nonsense, so can others: if IOMs are possible, then nonsense might be an *effective* excuse. Nevertheless, unless Hacker can justify his view that in this case no criterion means no sense, then, as Wilson goes on to suggest, he is surely using a stipulative notion of meaninglessness. I can find no attempt at such justification, not even in his scholarly chapter on the term 'criterion'.[34] Why does the fact, assuming it is a fact, that there is no criterion for the bottle's being there make it senseless to suppose it is? It is hardly ones immediate reaction to finding that one can think of no criterion to say, 'So I was wrong to think the

31. *An Analytical Commentary on the Philosophical Investigations*, Blackwell, 1990, Vol.III, p.408fn.

32. See 'Mental Events', in Donald Davidson *Essays in Actions and Events* Clarendon, 1980.

33. *Wittgenstein's Philosophical Investigations - A Guide*, Edinburgh U.P., 1998, p.44.

34. Op. cit., pp.545-68. There are two questions: is this way of using the notion of criterion faithful to Wittgenstein and, more importantly, is it justified?

suggestion made sense.' I am aware that Hacker would claim that allowing sense to the supposition involves treating mental images as more like physical pictures than they are, but surely one needs to evaluate the no-criterion argument *in order to decide* in what respects they are alike.

The kind of verificationism that pervades *Dreaming* is, it seems, not dead; so, even if the sorts of difficulties that beset Malcolm only arise with verificationist forms of nonsensicalism, this is of more than historical interest.[35] In fairness to Malcolm I ought to say that he would deny being a verificationist *tout court*. Elsewhere he writes:

> One cannot 'verify' that oneself feels hot, or hungry, or wants to sit down. It is ironic that some members of the Vienna Circle continued to insist that a sentence is meaningless if there is no possibility of verifying it, long after the originator of the Verification Principle [Wittgenstein] had abandoned it.[36]

Malcolm clearly agrees that it should be abandoned, but he seems only to be recognising quite specific exceptions to it (what are often called 'avowals'). The question remains: is it *ever* correct to reject a claim as meaningless because it is unverifiable? Or is it unverifiable because of what it means?

---

35. There is, of course, another area in which verificationist arguments are still heard - the realism versus anti-realism debate. But anti-realists often refrain from calling the unverifiable 'meaningless'. For an extended discussion of the complexities of the relationship between anti-realism and the question of meaningfulness see Bede Rundle, *Wittgenstein and Contemporary Philosophy of Language*, Blackwell, 1990, Chapters XI and XII, pp.228-9 especially.

36. *Wittgenstein: Nothing Is Hidden*, Blackwell, 1986, p.136

I n the last chapter we saw that Malcolm's way of talking about nonsense faces certain difficulties of which he does not appear to be fully aware. I now wish to present them more generally as ones facing any philosopher who accuses another of talking nonsense.

Specifying What One is Calling 'Nonsense'.

I begin with a problem that has been noticed by nonsensicalists and indeed is a recurring theme in the work of Wittgenstein. Consider the early Wittgenstein's reflexions on the question raised by Russell's Theory of Types of how to *prohibit* nonsense.

The theory says that certain types of symbol cannot be combined on pain of producing nonsense, e.g. 'The class of men is a man'. But it is not clear that this works. For what is meant by the following sentence? -

'The class of men is a man' is nonsense.

Kenny comments:

> If what is in quotation marks is meant to be just the sounds, or marks on the paper, then the whole sentence states at best a trivial empirical fact about arbitrary linguistic conventions, for there is nothing in that set of sounds which disqualifies it from being given a meaning.[1]

But, continues Kenny, Russell was concerned not merely with sounds or marks but with their meaning. So let us try:

> 'The class of men is a man', when that expression has the meaning it has in English, is nonsense.

This cannot be right, for if it is nonsense, it has no meaning in English. Should we then say then that 'the class of men' and '...is a man' cannot be combined so as to make sense? No, because if the expressions in quotation marks just refer to sounds, we are back with a trivial linguistic fact, whereas if they refer to sounds with their meanings, we must ask what is meant by 'combine' when we say they cannot be combined. Kenny comments:

> The most plausible account is: 'the class of men', meaning what it does in English, cannot be the subject of a sentence whose predicate is '... is a man', meaning what *that* does in

---

1. Anthony Kenny, *Wittgenstein*, Allen Lane The Penguin Press, 1973, p.43.

English. We may doubt whether this is in turn meaningful, but even if it is, it may well only postpone the evil day. For can we account for the meaning of the fragmentary expressions without giving an account of the sentences in which they can occur? If not, all our earlier problems will meet us again.[2]

There is a difficulty therefore in picking out a piece of nonsense in order to reject it, if this is to amount to more than a recognition of the contingent fact that a particular combination of sounds or marks is not part of the language. Wittgenstein concluded that this is something we cannot *say*: the rejection can only be *shown* by the adoption of a suitable notation. We talk nonsense ourselves if we try to say that signs with such-and-such meanings cannot be combined and if we try to explain why these combinations are prohibited.[3]

This line of thought reappears in the later philosophy. Consider *PI* 500

> When a sentence is called senseless, it is not as it were its sense that is senseless. But a combination of words is being excluded from the language, withdrawn from circulation.

Again he seems to be warning us against trying to exclude nonsense by saying that something is nonsense *if it is meant in a certain way.*

The Problem of Specifying the Nonsense might seem on the face of it to concern only the enlightened: one has realised that something is nonsense and one wants to earmark it as such. One must find a way of doing this without, as it were, contaminating oneself with the nonsense, treating it as if it made sense. (One could call the problem the 'Contamination Problem'.) But why are philosophers concerned to reject locutions as nonsense? Surely because they think the nonsense has been mistaken for sense. As Kenny says, they are not in general interested in the trivial empirical fact that a certain combination of sounds or marks is not a sentence of a certain language. At *PG* p.130 Wittgenstein makes the same point as at *PI* 500 and continues:

> [certain combinations of words] are excluded from our language like some arbitrary noise, and the reason for their *explicit* exclusion can only be that *we are tempted* to confuse

---

2. Ibid., p.44.

3. Ibid., p.44; *NB* 116; *TLP* 3.33-3.331. The discussion in the next chapter of Cora Diamond's interpretation of Wittgenstein is also relevant.

them with a sentence of our language.

But then can we avoid talking about those who succumb to the temptation, about their confusion and how it occurs? One way to see this is to reflect on the idea that a well-chosen notation will prevent or help to prevent us from talking nonsense.

Suppose we have what we think is such a notation but a philosopher complains that it is defective because it does not allow him to say something he wants to say. We reply that this is because 'saying it' would be talking nonsense. How can the dispute be resolved? It may be clear that the notation does not enable him to say 'what he wants to say'. But could it be so well-designed that it reveals to him that he is talking nonsense, the very attempt to say 'what he wants to say' in the notation somehow unmasking the nonsense, showing that there is nothing he really means? The answer is not obvious. So, even if Wittgenstein's way out of Russell's difficulties is acceptable, it is unclear that a philosopher who wants to accuse *someone else* of talking nonsense can avoid the problem of specifying the nonsense and saying why it is nonsense.[4]

But the fact that the philosopher's concern with nonsense tends to involve accusing *others* of talking nonsense leads to what seems to me to be a far more obvious problem.

The Problem Of Diagnosis: How Could One Ever Know That Someone Was Mistaken In Thinking He Meant Anything?

The Problem of Specifying the Nonsense has at least been noticed. This next problem is one that I have never seen clearly stated, at least so as to give it its full generality. It is this: how can A who accuses B of talking nonsense ever know that that is what B is doing? A might not understand B, of course, but that is not enough. Indeed nonsensicalists are in a sense claiming to understand those they accuse of talking nonsense better than they understand themselves. It is possible that what B says is not an acceptable sentence of the language, but that is also not enough unless A can show that no sentence in the language could express what B means - because he means nothing. (I take it that no one is going to claim that one cannot mean anything by a deviant sentence.) A cannot ascribe a meaning to what B says and on the basis of that show that he means nothing - as

---

4. The difficulty is exacerbated if one believes, as did the later Wittgenstein, that one can utter what *is* a possible sentence of the language and yet be talking nonsense, i.e. not mean anything by it oneself, even though one thinks one does.

distinct from showing that what he says is false. Once one has ascribed a meaning to something, one can no longer denigrate it as literally nonsense. One can say, 'If you mean that, you are wrong', but not, 'If you mean that, you don't mean anything'. But how can one get to work on someone's utterance without ascribing meaning to it?

The difficulty is so closely related to the last that it is surprising that it is almost universally ignored. Philosophers who seem well aware that one cannot say of a senseless locution that it is 'its sense that is senseless' seem not to notice that one cannot say that *what someone means by it* is meaningless. Yet once this has been noticed, it is not clear what line of attack is open to the nonsensicalist. Sometimes he can truly say that he does not understand what the other says; sometimes that what the other says is not a possible sentence of the language; sometimes both. But can he do more than this? He cannot make any assumption about whether the other means anything; for it would be self-defeating to accord a meaning to what the other says and question-begging to assume that he means nothing. So how is he to proceed?[5] It might be suggested that one can get a grip on what someone is *trying to do* or *thinks he is doing* when he comes out with an utterance, that there is a way of understanding *him* which does not require understanding *what he says*. This will need to be investigated. Here I just want to stress that there is a problem for nonsensicalists which they need to recognise.

The problem has in fact been noticed in connexion with verificationism. It has been pointed out that (as we found out with Malcolm) one seems to have to ascribe meaning to an utterance in order to decide whether it is verifiable.[6] This is especially clear if it requires some argument to reach a decision. Take theism: not everyone would consider it immediately obvious that the claim that God exists is unverifiable. (The traditional Argument from Design attempts to prove the claim empirically.) So one can say that if it is unverifiable, argument will be needed to show this. Would it then be possible to add that its unverifiability showed it to be meaningless? Significantly, the positivists often hedged their accusations by saying

5. I hope it is clear that the difficulty is not *just* that of proving the truth of a negative existential proposition, though the fact that we are dealing with such a proposition ('There is nothing he means.') ought to make us wary.

6. See, for example, Morris Lazerowitz, *The Structure of Metaphysics*, Routledge and Kegan Paul, 1955, pp.54-6.

that the claims they attacked were 'factually', 'cognitively' or 'empirically' meaningless.[7] One wonders whether this was to do more than invent new synonyms for 'unverifiable'. Be that as it may, it was no longer clear that someone making an unverifiable claim was being accused of meaning nothing whatsoever, in any sense of the word 'mean'.

The Inference Problem.

We saw when discussing Malcolm that there is a problem about how the nonsensicalist is to avoid treating what he claims is nonsense as something that can stand in logical relations with other locutions. He cannot, for example, refute a claim by inferring something from it which he then claims is nonsense. A confusion with *Reductio Ad Absurdum* may have helped obscure this point. It is of course legitimate to refute a claim by inferring from it something that is false. If it is patently false, then it can correctly be called 'absurd' and so can genuine nonsense. But it is necessary to insist that nonsense cannot entail or be entailed by anything.[8] Nor is it helpful to talk about what something *would* entail *if* it were meaningful. What it entailed would depend upon what it meant; if it now means nothing, it could be given any meaning whatsoever.

Wittgenstein talks of passing from 'disguised' to 'patent' nonsense (*PI* 464). Elsewhere he talks, somewhat mysteriously, about 'operations' that are needed to effect this.[9] Whatever these operations are, they cannot be, in any straightforward sense, *logical* operations: they cannot involve the replacement of the supposedly nonsensical expression by a synonymous one or the making of inferences from it.

It is worth noting that there is also a problem about what the nonsensicalist should say about the apparent inferences made by

7. See, for example, A.J. Ayer, *Language, Truth and Logic*, Gollancz, second edition, 1946, Chapter One.

8. Compare Edward Witherspoon, 'Conceptions of Nonsense in Carnap and Wittgenstein' in *The New Wittgenstein*, ed. Alice Crary and Rupert Read, Routledge, 2000, p.348 fn.34.

9. *Wittgenstein's Lectures, Cambridge 1930-32, from the notes of A. Ambrose and M. Macdonald*, ed. A. Ambrose, Blackwell, 1979, p.64.

10. It is gradually emerging just how insidious and convincing IOMs are going to have to be. Everything associated with genuinely meaningful discourse - inference, inconsistency, presupposition, clarification, disambiguation, the careful weighing of one's words - will have its illusory counterpart.

those he accuses of talking nonsense. Clearly he has to hold that sometimes they think they are inferring '*p*' from '*q*' when either '*p*' or '*q*' (or both) is nonsensical.[10] He needs to give an account of how such an error is possible.[11] I suppose he will have to give some kind of formalist account: the error involves the mere manipulation of sounds or marks on paper but the manipulator wrongly thinks he is taking cognizance of *meanings* that the sounds or marks have. Whether such an explanation would be plausible is hard to say in advance of knowing how, if at all, IOMs are possible.

Some Further Traps For The Nonsensicalist.

These are the main ways in which the nonsensicalist runs the risk of treating nonsense as if it made sense. But we found several other ways in which Malcolm made this mistake. I shall consider them briefly and argue that here too there are dangers for nonsensicalists in general.

In discussing dream-scepticism Malcolm seemed to think he could adopt 'a nonsensical assumption' for the sake of argument. One can see how this is going to be a temptation for the nonsensicalist. Philosophers are apt to attack their opponents not just with single arguments but with whole batteries of arguments. This is often legitimate but when nonsense is in question it is well to remember that if an utterance really is nonsensical, this is in an important sense the *only* criticism that can be made of it. If a nonsensicalist is tempted to pile on further criticisms—'Not only is it nonsense but it contradicts what you say about ...; it is open to the objection that ...; even if it made sense, it still wouldn't explain ..., etc.'—he is confused.

I wondered at one point whether Malcolm thought he could attribute 'nonsensical assumptions' to others, just as one can attribute false assumptions, even deeply buried ones, to them. Whether or not Malcolm is guilty of this, might it be a temptation for other nonsensicalists? Consider an example. It is common to say that philosophers such as Descartes and the British Empiricists are committed to the possibility of a private language. But notoriously none of them ever mentions 'private languages'; the assumption is

---

11. He is in fact claiming to have discovered a new kind of fallacy. Suppose someone 'infers' from '*p*', which makes sense, '*q*', which does not. A valid inference is truth-preserving. This 'inference' is not even meaningfulness-preserving; *a fortiori* it is not truth-preserving. It seems reasonable to call it a fallacy.

not one that they make explicitly. So it must be shown by argument that what they do say commits them to it.[12] There is no problem so long as the claim that there can be a private language is regarded as meaningful, though perhaps false, even necessarily false. But many writers on this topic clearly regard it as nonsensical. But can one 'make nonsensical assumptions', especially deeply buried ones? If not, how should the writers have put the point? (Would it be sufficient to say that Descartes and others are making the *false* assumption that a putative private language user would be speaking meaningfully?) This problem is of course linked to the Inference Problem but it seems worthy of separate mention.

When discussing Malcolm's first objection to the coherence principle - that one might dream that one's present experience cohered with one's past - I pointed out that, though Malcolm ought to consider his own objection nonsensical, to others it ought to appear ambiguous. Without claiming to know of a precise analogue to Malcolm's predicament in other nonsensicalists, I would argue that there is a trap for them which one might be tempted to call 'The Problem of Ambiguous Nonsense'. Suppose a philosopher writes an article attacking, say, the Mind-Brain Identity Theory but specifies that he is only targeting certain versions - others might be defensible. So far there is no difficulty: philosophers often narrow their sights in this way. But suppose it turns out that he thinks the versions he is attacking are nonsensical. He appears to be distinguishing 'nonsensical interpretations' of a theory from ones that may make sense and indeed various 'nonsensical interpretations' from each other. He can of course refer to the texts of the Identity Theorists themselves, but if he finds himself saying, 'Taken this way, A's thesis makes sense but taken that way, it doesn't' or 'There are two possible interpretations of what B says, neither of which makes sense', he had better think again.[13] What should he say? If IOMs are possible, there seems no reason why someone should not see an illusory meaning, or even more than one, in something that has a genuine meaning. On the other hand, it will not be easy for the nonsensicalist to discuss cases of this sort without contaminating himself with the nonsense.

Nonsense is not a species of falsehood; not an inferior kind of sense; not something that can be meant or understood. When a nonsensicalist like Malcolm forgets this and talks as though nonsense were not very different from sense, we must ask whether

40

this is merely a lack of rigour which could in principle be corrected or a pointer to something deeply wrong, something self-defeating, in nonsensicalism. Nonsensicalists must find ways of talking about, alluding to, nonsense and nonsense-talking that do not involve ascribing sense to the supposed nonsense. Nonsensicalism, one might say, imposes new obligations on philosophers, ones for which their traditional concern with truth versus falsehood does not well prepare them.[14] And even if they do devise ways of paying more than lip-service to the fact that it is nonsense they are supposed to be talking about, it is, I shall be arguing, still significant that they are constantly tempted to talk about nonsense as if it made sense.

---

12. See Kenny, op.cit., pp.16-17 for a discussion of the sense in which Descartes and others may be said to assume the possibility of a private language.

13. In *Insight and Illusion: Themes in the Philosophy of Wittgenstein*, Clarendon Press, revised edition, 1986, p.287, Hacker writes:

> ...if one means that someone else cannot think my thoughts, that is either false or nonsense. It is false if it means that another cannot think the very same thought or that he can never know what I think (even if I tell him!); it is nonsense if it means that another cannot 'do' my thinking.

14. Wittgenstein once wrote:

> I myself still find my way of philosophising new, & it keeps striking me so afresh, & that is why I have to repeat myself so often. It will have become part of the flesh & blood of a new generation & it will find the repetitions boring. For me they are necessary. - The method consists essentially in leaving the question of *truth* and asking about *sense* instead. (*CV* p.3)

Perhaps Wittgenstein is anxious not to forget to raise the question of sense with regard to philosophical issues - not to slip back into treating them in terms of truth versus falsehood. My point is that the nonsensicalist should not discuss the meaningfulness of locutions *as if* he were still discussing their truth or falsity. (Of course, if a locution is nonsense, it is *true* that it is nonsense and *false* that it makes sense.)

Chapter Four: Are Philosophers Who Make Accusations Of Talking Nonsense Really Postulating IOMs?

Perhaps it will be suggested that the difficulties I claim confront nonsensicalists only seem to do so because I am imposing on them an excessively strict conception of what nonsense and talking nonsense are. Are they in fact assuming that those they accuse are subject to IOMs? Obviously, there is no question of showing that every philosophical accusation of talking nonsense that has ever been made is nonsensicalist in my sense. What I shall do is to spell out in detail what the idea of IOMs amounts to and argue that any attempt to drop or water down any of its implications leads to something that either has no great philosophical significance or has only a loose connexion with nonsense (strictly so called), the distinctive feature of much Twentieth Century philosophy, its focus on meaning as prior to truth, being thereby lost.

What Is Involved in The Notion Of IOMs?

There are five main points:

a) Colloquially, 'nonsense' is often used to refer to what the speaker believes is manifestly false. This may indeed be its commonest use. Clearly someone talking nonsense in this sense means something. I think that philosophers normally avoid this use or at least that they have done so since accusations of talking nonsense became common.

b) When a philosopher is accused of talking nonsense it is not usually implied that he is doing so deliberately. Occasionally accusations of charlatanry are made, Schopenhauer's comments on Hegel being an early example. But I am not sure how often it has been seriously alleged that a philosopher has deliberately tried to dupe others with volumes of meaningless verbiage. There would in any case be a problem about what to say of those claiming to understand him.

c) When one philosopher accuses another of talking nonsense, he does not just mean that he does not understand him. This can be obscured by the fact that some philosophers express their criticisms in a semi-autobiographical form, documenting their own failure to understand a certain 'claim'.[1] But this is surely intended to show more than their own incapacity. It is supposed to help the reader see that there is something wrong with the 'claim'. (Philosophical criticisms often take the form, 'If he means such-and-such, I object that ...; if he doesn't, I don't know what he means.' But these are not nonsensicalist but straight criticisms of the other's utterances, interpreted in the only way that occurs to one.) Sometimes, of course, one simply has no idea what someone else means, so that the only criticism - if that is the word - that one can make is, 'I don't understand'. This can happen to a nonsensicalist as easily as to anyone else and it encourages the suspicion that he has to attach some sense to an utterance even to attempt an assessment of it.

d) When a philosopher is accused of talking nonsense, the charge is not just that what he says is not a possible sentence of the language. The charge might have been prompted by something aberrant in the wording of the utterance but the question will then be whether it can be reformulated in a way acceptable to the accuser. If it can, the accusation will have failed. It is easy to produce what is in a sense nonsense unintentionally—by grammatical errors, malapropisms, spelling errors, and so on—but in such cases the mistake is usually easily rectified and it is clear *that something was meant.*[2] In fact, as we shall see, it is the view of the later Wittgenstein that someone can produce an acceptable sentence of the language, one that has a use in the right circumstances, and yet be talking nonsense: he does not mean anything by it because he is not using it in the right sort of circumstances and has not specified a new use for it. On this view, producing an unacceptable sentence is not only not a sufficient condition for talking nonsense, it is not a necessary condition either.[3]

e) The accused believes *that he himself means something* by the locution he utters, not just *that it has meaning.* One can think that a locution has meaning when it does not without claiming to mean anything by it oneself: one might, for example, be duped into thinking that something is an Italian sentence when it is merely a random jumble of Italian words or Italian-like sounds. Whether or

---

1. There are several examples of this in Peter Winch's *Trying To Make Sense*, Blackwell, 1987.

2. Conversely, most locutions that are attacked as philosophical nonsense do not in any obvious way offend against what are ordinarily taken to be the rules of grammar and usage. We have already considered 'I am asleep' and 'I am dreaming'.

3. Arguably all nonsensicalists are committed to the view that someone can be wrong to think he means anything by a locution even though that locution *is* a possible sentence of the language, but it has to be admitted that the point only emerges with any clarity in the writings of the later Wittgenstein.

not such an error occurs much in philosophy, it is clearly possible. But what is almost always at issue when a philosopher is accused of talking nonsense is whether *he* means anything: he claims he does, whereas his accuser claims he is mistaken.

Points (d) and (e) can conveniently be expressed in terms of the distinction between 'speaker's meaning' and 'word-meaning':

d') A nonsensicalist accusation is not simply - or even necessarily - that what is said lacks word-meaning but that there is no speaker's meaning. (One might say that it is not about *word-nonsense* but about *speaker's nonsense*.)

e') The accused believes in the existence of a speaker's meaning, his own, not just that his utterance has word-meaning.

We have been concerned so far with the utterer of the alleged nonsense, but the nonsensicalist is not solely concerned with him. If the nonsensicalist thinks scepticism is nonsense, for example, he will regard as subject to IOMs not only those who voice sceptical doubts but those who claim to understand them (and perhaps think they can be *answered*). It is therefore necessary to add that an IOM might consist in someone's thinking he understands something by a locution when he understands nothing by it. Notice that it would have been wrong to say, ' ... someone's thinking he understands a locution *when he does not* '. This would have included *misunderstanding* the locution, thinking it (or its utterer) means one thing by it when it (or its utterer) means something else. That is clearly possible but it is not what the nonsensicalist has in mind. We need a term - say '*pseudo*-understanding' - to differentiate the alleged error of thinking one understands something by a locution when one understands nothing by it from merely *mis*understanding it.

I shall frequently speak of 'talking nonsense' and 'IOMs' to refer to the error the nonsensicalist postulates. It would drive the reader to distraction if every time I were to spell out what this involves. Nevertheless I shall also frequently use such longer formulations as 'mistakenly thinking there is something one means by what one says', 'being mistaken in thinking one means anything' and 'thinking one understands something by a locution when one understands nothing by it'.[4] This is because I believe that nonsensicalists often forget precisely what is involved in their accusations or at least use terminology which confuses the supposed error in question with others that are clearly possible and unproblematic. For example, it is often not made clear that we are not just concerned with whether what is said is a possible sentence of the language. (We saw how Malcolm was often less than explicit on this point.) And philosophers use ambiguous turns of phrase such as, 'We wrongly think it makes sense to say (or ask) ...' which do not make it clear whether we are wrongly thinking a locution has meaning or that we mean something by it ourselves.[5] And misunderstanding and pseudo-understanding are sometimes not distinguished.[6]

Another terminological matter may conveniently be mentioned here. Most of the references to nonsense in this book will be to *alleged* nonsense, utterances that may or may not really be nonsense. And many of the references to claims, propositions, questions, and so on, will be to *putative* claims, propositions and questions, ones that some philosophers will deny are genuinely meaningful uses of language. If I were to be utterly rigorous, it is likely that every other sentence would contain a word like 'alleged', 'supposed' or 'putative', or some expression in scare-quotes. Here again I appeal to the common sense of the reader. Whilst I shall frequently include such words and devices as reminders, it surely ought usually to be obvious, given the subject-matter of the book, what the reader is expected to treat as problematic.

I believe that those philosophers who accuse others of talking

---

4. One turn of phrase I shall try to avoid is 'wrongly thinking one means *something* by a locution'. This is ambiguous: as well as the supposed error of being wrong to think one means *anything* it could refer to a supposed error of being wrong about *what one means*. If I have let the odd example creep into the text, I ask the reader's pardon but would remind him that since the book is almost entirely concerned with the first error I am likely to be referring to that. Later in this chapter I briefly consider an error which, if it is possible, will be of the second type.

5. For example, Hacker and Baker, *Analytical Commentary on the Philosophical Investigations*, Vol. II. Blackwell, 1985, p.21: 'So we assume ... that what it makes sense to say about seeing a tree (e.g. "I didn't notice any birds' nests, but there may have been some.") must make sense about imagining a tree' and (also p.21) 'It makes sense to ask whether there are four successive 7s in the first thousand places of the expansion of π. So we misguidedly think that it makes sense to ask whether there are four successive 7s in the expansion of π.'

6. For example, P.T. Geach, *God and the Soul*, Routledge and Kegan Paul, 1969, p.5: 'People often claim in all good faith to understand, and feel as though they understood, and yet turn out not to have understood; anyone who has to teach philosophy becomes well aware of this if he marks his own pupils' papers. And some sorts of pseudo-supposition give us a strong feeling of understanding in spite of gross incoherence.' The reader may need to look up the context to satisfy himself of this, but Geach seems to be adducing students' *mis*understandings as evidence for the possibility of *pseudo*-understanding.

nonsense have normally meant and been taken to mean that the accused is wrong to think he means anything and I do not expect nonsensicalists to protest, 'We never said that!' One might wonder, however, whether there could be other ways of taking such accusations and - as already suggested - whether those who make them have always borne clearly in mind what they involve. And I suspect that unclarity about what is involved might have led some philosophers to think that it is just obvious that if someone said certain things, he would be talking nonsense. Let us consider a possible case.

What Time Is It On The Sun?

I have known a philosopher claim that someone who asked this would be talking nonsense. When I admitted that there was *something* wrong with this question or 'question', he seemed to think that I was pretty well conceding the point at issue. This seems to me to come dangerously close to using the word 'nonsense' as a catch-all for any utterance that embodies an error of some sort but which could not simply be described as false. (Many philosophers seem to use the word 'incoherent' in a similar unhelpfully vague way.) If he had meant that someone asking the question must be mistaken in thinking he meant anything by it, then he would have been making the kind of claim with which I am concerned. The issue between us would have been sufficiently clear-cut for discussion to proceed.

Or perhaps he thought there was more than one possibility here. He might have admitted that someone who said, 'I believe that conventions have now been established for telling time on the sun. What, according to them, is the time now?' would have been asking a perfectly meaningful question based on a clearly stated false assumption.[7] But he might have claimed that if the speaker had simply not realised that there is no obvious or natural way of extending our convention for telling time at a particular place on the earth to the sun and that therefore something novel would be required, then he would have been talking nonsense in the sense of thinking he meant something when he meant nothing.

---

7. As I am recalling an actual episode, I have retained the reference to 'time on the sun' (clearly inspired by *PI* 350-1), but I think that in many ways 'time at the North Pole' would have furnished a better example. It would have been much easier to imagine confusions and misunderstandings that might occur in real life (and have practical consequences) and to say with some precision what would probably have been going on in the minds of those committing the errors.

I draw two morals from this case. First, doubts about nonsensicalism cannot simply be laid to rest by adducing examples. Certainly it is easy to find or imagine instances of people saying bizarre things. But nothing follows about whether they are talking nonsense. If one thinks they are suffering from IOMs, one must explain how IOMs are possible and how one could know that someone was suffering from one. There is no such thing as pointing to clear and obvious examples of IOMs: they are always something postulated. I shall expand this point in the next chapter.

The second moral returns us to the question of whether I am right about what philosophers who make accusations of talking nonsense are actually saying. I said that someone who thought that conventions already existed for telling time on the sun and wanted to know what, according to them, was the time now would not be talking nonsense. Perhaps it will be felt that it is my conception of nonsense that is too narrow. Could we not describe questions based on false assumptions as nonsense?[8] It has to be admitted that there is no standard way in non-technical English of referring to questions based on false assumptions. One certainly does not call them 'false'. So perhaps one could call them 'nonsense'.

But would this be a sensible terminology to adopt? What would nonsense of this sort really have in common with nonsense strictly so called, where an absence of meaning is implied? Would it not be, at best, an extension of the colloquial use of 'nonsense' to refer to the patently false? Given that the asker of the question means something by it and has expressed himself perfectly adequately, is it not positively misleading to call it 'nonsense'?[9] The intention behind

---

8. Especially, it might be thought, since the false assumption concerns the existence of (linguistic) conventions. I am not sure though how much difference this makes. Suppose I ask for the postcode of a foreign correspondent and it turns out that in his country there is no equivalent of the British system of postcodes. Could one really call my question nonsense, even though for it to have a *direct* answer conventions would need to be in place, conventions that are in a broad sense linguistic?

9. Even if the speaker had forgotten or were unaware that our conventions governing local time on the earth depend upon the position of the sun in the sky, why could one not treat his question as a perfectly meaningful one to which the answer is, 'There is no recognised local time on the sun because ...'? This no more treats it as nonsense than answering 'There is no bus from here to X' to 'When is the next bus to X?' treats that as nonsense. I sometimes wonder whether there is a danger of succumbing to illusions of *meaninglessness* rather than of *meaning*. Certainly questions and statements that presuppose falsehoods often seem *odder* than straightforward falsehoods (as though they were somehow even more remote from reality) and many philosophers are quick to move from oddity to nonsensicality. But if the question or statement is rephrased so that what is presupposed is stated explicitly, the temptation to do this is weakened, often to vanishing point.

philosophical accusations of talking nonsense is surely to effect a change of emphasis: to say that questions about meaning are prior to questions about truth and to stress the alleged possibility that we may think we are arguing about the truth of a proposition when we have not yet given it a meaning.[10] So my second moral is really a recommendation to the nonsensicalist: even if there is a case for calling a certain error 'nonsense', you should ask yourself whether it is really what you had in mind when you introduced accusations of talking nonsense into philosophy.

Might There Be 'Meanable' Or 'Thinkable' Nonsense?
One reason one might have lingering doubts about whether philosophers who accuse others of talking nonsense are really postulating IOMs is that it is surprisingly difficult to find explicit statements about what *talking nonsense* involves though there is much more on what *nonsense* (the sounds or marks on paper) is.

It is possible to find some examples of philosophers who accuse others of being mistaken to think they mean anything. Wittgenstein's *Tractatus* claim that a would-be metaphysician needs to be shown that he has given no meaning to certain signs in his propositions (6.53) is one example and Anscombe once remarked of Flew that he 'did not really mean anything (even if he felt as if he did)'.[11] And in Chapter Seven I shall be considering what has been done to show how such errors might be possible. Furthermore, I do not know of a single case where a philosopher tries to tone down an accusation of talking nonsense by saying that he does not mean that the accused is wrong to think he means *anything*. Nevertheless it might be objected that I have at most shown that some philosophers have some of the time meant by their accusations of talking nonsense that the accused was mistaken in thinking he meant anything. Is there any other way of taking them? There is one possibility or seeming-possibility that

---

10. It is possible to go quite a long way with the nonsensicalist here without actually becoming a nonsensicalist. One might agree that there are questions that manifest a certain naivety about the prospects for answering them *just as they stand* without calling them nonsense and assuming that those who ask them mean nothing.

11. G.E.M. Anscombe, *An Introduction to Wittgenstein's Tractatus*, Hutchinson, 1959, p.85.

does need to be discussed. Could it be that sometimes what is intended is not that the accused does not mean anything but that *what he means is nonsensical*? Are we right in equating the nonsensical with the meaningless, more precisely equating talking nonsense with meaning nothing by what one says?

Suppose someone is accused of talking nonsense. He protests that he does mean something and proceeds to explain what. His accuser interrupts, 'You misunderstand. I did not say you meant nothing. I said you were talking nonsense. What you *meant* was nonsense.' Now he might be using 'nonsense' in the colloquial sense. Or perhaps he is accusing the other of contradicting himself. If he means either of these things, it is hard to see that there is anything distinctive about the accusation he is making: we all reject claims we think are manifestly false or inconsistent. But is there any other possibility?

At *TLP* 5.5422 Wittgenstein complained that a theory of Judgment of Russell's did not make it impossible to judge a nonsense. If he did not think it possible to *judge* nonsense, it is unlikely that he would have thought it possible to *mean* nonsense. And in his later work he maintained that to say that a locution is senseless is not to say that its sense is senseless (*PI* 500). He did not have in mind some sense of 'senseless' that is compatible with meaningfulness.[12] It might be objected that *PI* 500 is concerned solely with combinations of words, whereas we have been concerned with whether *a person* means anything by such combinations. This is true, but if he had believed it possible for someone to mean a nonsense by a locution, it is hard to see how he could have avoided allowing that a locution can have a meaning and yet be nonsensical: all that would be necessary would be that some group of speakers should mean the same nonsensical thing by some locution.

It looks therefore as though Wittgenstein would not have allowed the possibility of 'meanable' or 'thinkable' nonsense. But was he right in this? And what of other nonsensicalists? Might it be held that the pathology of thought parallels the pathology of language? If there can be nonsensical combinations of words, why not nonsensical combinations of thoughts (or perhaps thought-constituents)? Consider another passage from Wittgenstein:

---

12. It is important to remember that 'senseless' is not contrasted with 'nonsensical' in the *Investigations* as it is in the *Tractatus. PI* 500 can be taken to be about locutions which are senseless, meaningless or nonsensical.

DO PHILOSOPHERS TALK NONSENSE?

When we say that a thing cannot be green and yellow at the same time we are excluding something but what? Were we to find something which we described as green and yellow, we would immediately say that this was not an excluded case. We have not excluded any case at all, but rather the use of an expression. And what we exclude has no semblance of sense. Most of us think that there is nonsense which makes sense and nonsense which does not - that it is nonsense in a different way to say, 'This is green and yellow at the same time' from saying 'Ab sur ah'. But these are nonsense in the same sense, the only difference being in the jingle of the words.[13]

This is a clear rejection of the idea of meaningful nonsense. But it also shows just how uncompromising his view is. Most of us probably think that 'This is green and yellow all over at the same time' means *something* (though we may think it necessarily false). If we become nonsensicalists and dismiss it as nonsense, we are likely to do so in a half-hearted fashion, thinking of it as not quite as meaningless as 'Ab sur ah'. Wittgenstein will have none of this. Something is either meaningful or it is not; we have either given it a meaning or we have not; we either mean something by it or we do not.

Cora Diamond in 'What Nonsense Might Be'[14] presents an interpretation of Wittgenstein on nonsense that is closely related to that given here. She distinguishes what she calls the 'natural view' of nonsense from that of Wittgenstein (both early and late). Consider the two locutions:

(i) In the next room there is a splurg.

(ii) In the next room there is a however.

The natural view is that both are nonsensical but in different ways: (i) because no meaning is standardly given in English to 'splurg', (ii) because, although 'In the next room there is a ...' and 'however' are meaningful elements of English, they clash producing nonsense

---

13. *Wittgenstein's Lectures: Cambridge, 1932-35 from the notes of Alice Ambrose and Margaret Macdonald*, ed. Alice Ambrose, Blackwell, 1979, p.63. I think the sentence 'And what we exclude has no semblance of sense' has to be taken as an emphatic way of saying that what is excluded has no sense. If one tried to take it literally - as saying that the excluded does not even *seem* to make sense - it would go against the whole tenor of the passage, indeed against Wittgenstein's general nonsensicalism.

14. In *The Realistic Spirit: Wittgenstein, Philosophy and the Mind*, MIT Press, 1991. Here I present only her conclusions, not her evidence.

when combined. On Wittgenstein's view (which she favours) they are nonsensical in the same way, for no meaning has been given in English to 'however' as a noun referring to an object with spatial location. (Equally we could say that English gives no meaning to 'In the next room there is a ...' preceding a word like 'however', which does not refer to an object with spatial location.) The fact that 'however' has a meaning in the right sort of context, whereas 'splurg' has no meaning in any context, is irrelevant. 'However' does not carry with it the meaning it has in the right sort of context when it is put in the wrong sort of context. There is nothing to prevent us giving either 'splurg' or 'however' a meaning suitable to the verbal context in question; it is just that this has not been done. On the natural view there is both 'negative nonsense' due to failure to give meaning to certain signs and 'positive nonsense' due to incompatibility of meanings. On Diamond's interpretation of Wittgenstein's view there is only 'negative nonsense':

> *Anything* that is nonsense is so merely because some determination of meaning has *not* been made; it is not nonsense as a logical result of determinations that *have* been made.[15]
>
> There is no 'positive nonsense', no such thing as nonsense that is nonsense on account of what it would have to mean, given the meanings already fixed for the terms it contains.[16]

Diamond is solely concerned with combinations of sounds or marks on paper, whereas we want to know about the person producing them. From our point of view her discussion of 'what nonsense might be' needs to be supplemented with a discussion of 'what *talking* nonsense might be'.[17] But the argument given earlier applies here: if it were possible to mean that in the next room there is a however (even though this is nonsense), then it would be possible for some group of English-speakers to mean that in the next room there is a however by 'In the next room there is a however', and we

---

15. Ibid., p.106.
16. Ibid., p. 107
17. Diamond's discussion does not seem immediately applicable to one kind of nonsense recognised by Wittgenstein: that where a philosopher utters a sentence that *is* meaningful, provided it is used in the right kind of situation. Clearly, the only way of giving an account of this sort of nonsense - assuming it exists - would be in terms of what it is to *talk* nonsense.

would have a sentence that was meaningful but nonsensical. Thus if one denies the possibility of positive nonsense, nonsense produced by combining elements which retain their meaning when combined, one seems committed to denying that someone could mean a nonsense.

Might it be that some philosophers have accepted nonsensicalism as easily as they have because they have (rightly or wrongly) not followed Wittgenstein in his uncompromising view of nonsense? Would all nonsensicalists happily admit that they are committed to regarding everything they dismiss as philosophical nonsense as having no more meaning then 'Ab sur ah'?[18] It is not that I believe that there is in contemporary philosophy some fully worked-out alternative view. It is rather that I think it at least possible that some philosophers have no precise account of nonsense (and talking nonsense) and so have taken up positions of which they might have been more wary if they had tried to be more precise. Each nonsensicalist must judge for himself.

Cora Diamond is one of a group of philosophers contributing to a recent volume of essays on Wittgenstein[19] who see him as taking this uncompromising - 'austere' is the favoured term - view of nonsense. They argue that many other philosophers have been insufficiently rigorous in their employment of the concept of nonsense and that has led them to misinterpret Wittgenstein[20] and to make various errors in their own philosophising. Edward Witherspoon, for example,

---

18. Hacker in *Insight and Illusion*, Clarendon Press, revised edition, 1986, p.18, writes, 'The claim that all past philosophy is riddled with error is quite common in the philosophical world. But that it is all a subtle form of gibberish is not.' One wants to protest that the latter claim was common enough in the Twentieth Century. But perhaps not all those who have made accusations of talking nonsense have thought through what this involves as clearly as Wittgenstein: they have not necessarily accepted the 'Ab sur ah' Principle.

19. *The New Wittgenstein*, ed. Alice Crary and Rupert Read, Routledge, 2000.

20. One question addressed by several of the contributors is whether the *Tractatus* tries somehow to convey insights about the relationship between language and the world which cannot be said (the usual interpretation) or whether Wittgenstein regarded the passages in which he seems to be attempting this to be nonsense and nothing more (the 'austere' theorists' interpretation). Much is made by the 'austere' theorists of the fact that in the famous 'self-condemnation' at 6.54 he says that his propositions serve as elucidations in that anyone who understands *him* (not *them*) recognises them as nonsense. Fortunately I need not take a stance on this issue here. I would suggest however that it is worth asking whether the difficulty of finding a stable and consistent interpretation of Wittgenstein on this point might be in part a reflexion of the difficulty of being a consistent nonsensicalist.

criticises Carnap, Hacker and Baker, and Marie McGinn for the latter.[21] What none of them seems to do is to entertain any doubts about nonsensicalism itself; they attack other nonsensicalists for treating nonsense as if it made a kind of sense but do not ask whether the frequency of this error points to any general problem with nonsensicalism. They assume that one can be rigorous about nonsense and still use the concept of nonsense as a philosophical weapon. I believe that I have already said enough to suggest that this cannot simply be taken for granted.

I do not find it easy to account for this reluctance to question nonsensicalism itself. In spite of the many insights in the anthology, some of which are mentioned in this book, there must be dozens of places where I think that here would be a good place to ask whether IOMs are possible or how they might be diagnosed but am disappointed. Even if the contributors think the answers obvious, they are surely not so obvious that they need never be given explicitly. Part of the reason for this neglect is no doubt the by now familiar tendency to say far more about nonsense than about *talking* nonsense. This encourages a concentration upon the Problem of Specifying the Nonsense rather than upon that of Diagnosis: there is a tendency to discuss cases where it has already been decided that something is nonsense and, I suspect, an underlying assumption that one can decide that something is nonsense *and only then* start being rigorous and austere in how one talks about it.[22] But this cannot be the whole story, for discussion of *talking* nonsense is not absent from these papers, still less from the writings of (the later) Wittgenstein. Perhaps one must recognise a general principle that blind spots are never easy to explain.

Some Questions About This Rigorous View Of Nonsense (And Nonsense-Talking).

There is one aspect of Wittgenstein's later philosophy that might not seem to fit the strict, uncompromising view of nonsense that I am

---

21. Opposed to the 'austere' conception of nonsense favoured by Wittgenstein and endorsed by these writers is the 'substantial' (see the essay by James Conant) or 'Carnapian' (see the essay by Edward Witherspoon) conception. In Diamond's terminology substantial or Carnapian conceptions allow for 'positive' as well as 'negative' nonsense.

22. This comment is less applicable to Witherspoon's paper than to some of the others. His suggestion about how one should go about diagnosing philosophical nonsense I consider in Chapter Nine.

attributing to him.[23] Commentators have noted a tendency of his to treat grammatical propositions, i.e. ones that really express grammatical rules, as nonsensical when their status is misunderstood. John V. Canfield quotes the following:

> [Wittgenstein] said that ... both the Realist and the Idealist were 'talking nonsense' in the particular sense in which 'nonsense is produced by trying to express by the use of language what ought to be embodied in the grammar'; and he illustrated this sense by saying that 'I can't feel his toothache' means ' "I feel his toothache" has no sense' and therefore does not 'express a fact' as 'I can't play chess' may do.[24]

He comments:

> ... Wittgenstein gives a clear account of how he uses the term 'nonsense' hereabout. Nonsense occurs if someone tries to assert as a ground-level claim in a language game something that is part of the grammar of the game. In general, Wittgenstein has a tendency, which survives to the *PI*, to restrict 'say', 'sense' and related words to what can be said as a contingent claim in some or another language game.[25]

Similarly J.F.M. Hunter:

> ... in a large number of places, Wittgenstein seems to take the view that grammatical points when not marked as such are nonsense if they sound like empirical generalizations, but not if their grammatical character is well understood.[26]

I agree that Wittgenstein does talk in this way[27] but wonder about

---

23. There might also be an aspect of the early philosophy that does not fit: his supposed attempt to insinuate ineffable truths by means of locutions that are strictly nonsense. This is precisely the interpretation that is rejected by some of the austere theorists (see fn.20). They see the early philosophy as being consistently austere.

24. G.E. Moore, 'Wittgenstein's Lectures: 1930-33', reprinted in *Ludwig Wittgenstein - Philosophical Occasions*, ed. James Klagge and Alfred Nordmann, Hackett, 1993, p.103.

25. *Wittgenstein: Language and the World*, University of Massachusetts Press, 1981, pp.156-7.

26. *Wittgenstein on Words as Instruments*, Edinburgh University Press, 1990, pp.16-17.

27. Other passages of interest in this connexion are those where, as in the passage quoted from Moore's report, he advocates the replacement of metaphysical claims about possibility with statements about what it makes sense to say (e.g. *Z* 134; *BB* pp.54-8) and where he discusses whether it is better to say, 'Of course' or 'Nonsense' to a grammatical proposition (*BB* p.30; *PI* 251-2; *PG* p.129).

his justification for doing so. He seems to admit that a grammatical proposition is not nonsense if the fact that it expresses a grammatical rule is clearly recognised. Thus it is not 'word-nonsense', a combination of words excluded from the language. We are assuming too that the rule is a genuine one, so that there is something right about the utterance, even if the speaker misconstrues the nature of what he has dimly perceived. This fits ill with the claim that he means or understands *nothing* by it. And the large claims that philosophers are apt to make for these supposedly misconstrued rules of grammar—that they are necessary truths, universally valid, and so on—seem only explicable on the assumption that such claims result from some kind of engagement with the meaning of the rules, even if this engagement involves viewing them through a distorting lens.

What seems to be implied is that in such cases philosophers are somehow wrong about the character of their own utterances. They treat grammatical propositions as propositions about what the world is like. They produce one kind of utterance, thinking they are producing another kind. It might be doubted whether such errors are possible. They seem to involve being wrong about *what* one means by one's own present utterance and it is not obvious that this is possible. But even if it is, why speak of 'nonsense'? What is said in such a case is not word-nonsense and it has to be assumed that the speaker means something by it. Even the colloquial use of the word 'nonsense' seems rather strained here: the speaker is making a rather sophisticated error, not uttering a blatant falsehood.

A way of bringing this use of 'nonsense' into line with Wittgenstein's generally rigorous use of the word might seem implicit in the above quotation from Moore's report. Could it not be argued that someone who misconstrues a grammatical rule, such as 'I can't feel his toothache', as a statement about the world must think he means something by its negation, 'I can feel his toothache', though in fact he means nothing by it? Unfortunately, it is not clear that this follows: all that may follow is that he must think it *has* an intelligible negation, not that he must think that he himself understands its negation. Perhaps not even this much follows: perhaps all that follows is that if he were to draw out the consequences of his mistaken view, he would think it had an intelligible negation.

A better way of dealing with the difficulty may be this. Perhaps Wittgenstein would say that someone who thinks that 'I can't feel his toothache' is a statement about the world, embodying some kind of universal necessary truth about the relation between people and their pains, does not just misclassify it—he seems to see a sense in it appropriate to his misclassification. There is no such sense and hence the philosopher is the victim of an IOM. There would be an obvious problem about whether he seems to see the illusory sense *in addition to* the genuine grammatical sense or *instead of* it. Clearly the genuine sense must somehow be responsible for the illusory one, but that does not answer our question. I suppose it would be best to say that he distortedly sees the genuine grammatical sense as a necessary truth about the nature of pain. If it is thought that we are riding the perceptual analogy too hard here, some such word as 'registers' could be substituted for 'sees'. (In fact, a disanalogy between perceptual illusions and what IOMs would have to be is now coming into view. One can specify what the victim of a perceptual illusion thinks he perceives, but one could not simply state what the victim of an IOM thought he meant. Or so I shall argue in Chapter Seven. Here the illusory sense has to be referred to indirectly by such phrases as 'universal truths about the world' or 'necessary truths about pains'.)

That is the best I can do by way of reconciling Wittgenstein's use of the word 'nonsense' in connexion with misconstrued grammatical propositions with the 'austere' conception of nonsense he develops elsewhere. But even if my attempt is found unconvincing, I do not think the 'austere' conception should be discounted. It surely represents his considered view of nonsense; but it may be that he has carelessly called 'nonsense' another kind of error which he, rightly or wrongly, believes is possible and which after all concerns meaning.

But is he right to take this rigorous view of nonsense? We have seen that he denies that 'there is nonsense which makes sense'. Could a case be made for the opposite view? Probably no one would boldly assert that there *can* be nonsense which makes sense: that sounds too much like a contradiction. But one might wonder whether the distinction on which Wittgenstein seems to be relying between making sense and not doing so is as sharp as he implies. Consider in particular speaker's meaning. Is there always a definite answer to the question whether someone means anything by an utterance? Can one say, 'Either he means something by what he says or he doesn't'? Suppose there were some vagueness here. One might be led to suspect that there is by considering some of the more bizarre cases of 'secondary sense' that Wittgenstein discusses in various places[28], locutions like 'Is Wednesday a fat or a lean day?' and 'Arrange the vowels in order of their darkness.'

I think the question of vagueness is important but I shall not pursue it here.[29] This is because it seems to me that if the charge of talking nonsense is to stick, the accused must *definitely* not mean anything: there must be no unclarity or indeterminacy. If one is dealing with a borderline case of meaningful speech (assuming such cares exist), then there must *ex hypothesi* be *something* to be said for treating it as meaningful; so one might as well treat it as such, thereby avoiding all the difficulties facing nonsensicalism, and ask what is to be said for or against it when it is so treated. One is, as it were, back in the realm of traditional, pre-nonsensicalist philosophy.

One other problem raised by the austere view of nonsense I shall discuss (inconclusively) in Chapter Six. Even if it is true that one cannot *mean* nonsense, does the same apply to all psychological verbs? If someone thinks, believes, imagines, desires, fears, ... that p, must 'p' always make sense? There is at least one verb - 'dream' - about which one might hesitate.

Falsidal Theories.

From time to time the view is put forward that much of what recent philosophers have stigmatised as nonsense should really be regarded as false. Such falsidal theories have been espoused by Ewing, Quine

---

28. *PI* Pt.II. p.216; *BrB* pp.135-41. 148; *Z* 185.

29. This is as good a place as any to mention Gareth Evans's view (*Varieties of Reference*. O.U.P.. 1982. p.338) that if someone who is hallucinating points into space and cries. 'He is coming to get me', he has failed to *say* anything. Not only that, there is nothing that he is *meant to say* and another person cannot understand what he says (though he can attain some kind of understanding of what is going on). This view has struck many philosophers as bizarre (e.g. Simon Blackburn. *Spreading the Word*, Clarendon Press. 1984. pp.318-19). Certainly, Evans seems to be imposing especially strict conditions on what it is to succeed in saying something and what it is for someone else to understand what is said. But could such stringency have any warrant from ordinary usage? And is Evans attributing some kind of IOM to his hallucinator? I shall not pursue these questions however since it is clear that they are remote from our concern with the denial of meaningfulness as a form of philosophical criticism. Philosophical nonsense is always held to arise from conceptual confusion and surely the most conceptually clear-headed person imaginable could be unfortunate enough to suffer from hallucinations. Evans's reasons for denying that someone else can understand the hallucinator's utterance centre on the claim that the putative understanding would require the acceptance of a *falsehood* (ibid.. p.331).

and many of the participants in a long-running dispute in the *Australasian Journal of Philosophy*.[30] How do such views relate to our problem? Not as straightforwardly as one might expect. For one thing, there is the tendency already encountered to discuss only the locutions and to neglect the person who utters them. It is therefore not entirely clear whether the adoption of falsidal theories commits the theorist to the rejection of the very possibility of IOMs.

But there is also, it seems to me, something artificial about this way of approaching the question of nonsense. Nonsensicalists claim that people sometimes think there is something they mean by certain locutions when there is not and that this happens frequently in philosophy. In the course of expounding their views they often invent locutions which seem superficially grammatical but which are supposed to be *obviously* nonsensical. But, instead of drawing the desired conclusion that we should be wary of the superficially grammatical, falsidal theorists reply that these locutions are meaningful but false. In one way this seems a shrewd response; but one wonders how much these locutions actually convey or seem to convey to those who take this line. Certainly they do not in general convey much to most people - which is why they are adduced. It may be that the techniques recommended by Michael Bradley, for example[31], can be used to show how 'Friday is in bed' or even 'Quadruplicity drinks procrastination' can be treated as false rather than meaningless, but ought one not, on pain of gross artificiality, to take some account of whether these locutions seem pre-theoretically to convey any meaning to anyone? What gives the claims of nonsensicalists their importance and bite is that they attack locutions that philosophers have traditionally seen as meaningful, as of great importance, and often as true not false. At any rate, it is the idea that one can be deceived in thinking there is something one means or understands by a locution that seems to me to be crucial for the philosophical concern with nonsense.

Clearly, if one rejects or has doubts about nonsensicalism, one will be faced with the question of what to say about the utterances that nonsensicalists have stigmatised as nonsense. But one needs to keep one's options open. The utterances may be true or false, but they may be neither - *questions* have been subject to nonsensicalist attack as often as have assertions. Often one may not know what to say. I myself do not know what if anything is wrong with sceptical doubts. I do not think they are nonsensical and so it seems to me that, if there is something wrong with them, falsehood will come in somewhere. But where or how I have no idea; if I had, I would probably be near to a final reckoning with scepticism!

30. A.C. Ewing, 'Meaninglessness' in *Non-Linguistic Philosophy*, George, Allen and Unwin, 1968, pp.15-33; W.V.O. Quine, *Word and Object*, MIT Press, 1960, p.229; Michael Bradley, 'On the Alleged Need for Nonsense', *Australasian Journal of Philosophy*, 1978, pp.203-18 (gives references to the other disputants).
31. Bradley, op. cit., pp.204-6.

Chapter Five: The Elusiveness Of Philosophical Nonsense.

I want now to discuss certain features of alleged philosophical nonsense which, though they do not prove there is no such thing as IOMs, do encourage scepticism about the notion.

The Staying Power Of Alleged Philosophical Nonsense.

When a philosopher proves to his own satisfaction that a locution is philosophical nonsense it does not normally lose all appearance of meaningfulness for him. Consider someone whose official view it is that scepticism is nonsense. He will almost certainly have no difficulty 'expounding' scepticism, explaining to beginners in philosophy for example, what it is and how it can seem plausible. This is hard to explain unless we assume that it still *seems* meaningful to him (though he thinks he knows it isn't). If sceptical utterances had come to seem no more meaningful to him than 'Ab sur ah', he would surely have had great difficulty expressing the sceptic's case: he might have been reduced to learning passages from Descartes and Hume by rote.[1] Compare this philosopher with one who in the ordinary sense does not understand what some other philosopher is saying. He would probably happily undertake to teach a course on epistemology, whereas someone who never claimed to understand Hegel, say, would be reluctant to lecture on Hegel.

The nonsensicalist has a ready reply. He will say that this just shows how *insidious* IOMs are and he might add:

> Are not *perceptual* illusions like this? Lines continue to look curved even though one knows they are straight or they look to have different lengths when one knows they have the same length. In fact perceptual illusions are typically producible to order: any normal person will experience them if conditions are right. Does this not parallel the fact that beginners in philosophy, even those of very modest ability, normally have

---

1. There are a number of empirical questions in the background which I shall not pursue here. Does it *ever* happen that a philosopher who comes to reject as nonsense a locution he once thought meaningful finds that it ceases even to *seem* meaningful? Does a locution which a philosopher has come to stigmatise as nonsense seem to him *in any way* different from when he thought it made sense or does he just pigeon-hole it differently? Are there examples of intelligent men to whom alleged philosophical nonsense - even such things as scepticism which beginners find it easy to (seem to) grasp - has never seemed meaningful? (From the nonsensicalist standpoint they would be immune to IOMs.)

no difficulty understanding (really: seeming to understand) scepticism? And even when they are disabused of the opinion that scepticism makes sense, it still *seems* to do so.

Perhaps. But this staying power of alleged philosophical nonsense does make one wonder whether it is nonsense at all: maybe one does not just *seem* to understand something by it, one really does. Everything will depend upon what reasons the nonsensicalist can adduce for dismissing appearances as misleading. There are familiar reasons why perceptual illusions are regarded as such; yet, although nonsensicalists do have their reasons for dismissing utterances as only seemingly meaningful, we shall see in the next section that no such dismissals are uncontroversial.

I should perhaps add that I am not accusing nonsensicalists of disingenuousness. Although we have probably all at times wondered whether another person (philosopher or non-philosopher) was merely *pretending* not to understand us, I am assuming that nonsensicalists genuinely believe they have grounds for dismissing the apparent meaningfulness of certain utterances as illusory. I am however suggesting that it may be these grounds that are at fault, not the utterances.

Wittgenstein recognises that locutions rejected as nonsense on philosophical grounds do not thereby cease to seem meaningful. However, I think he gives the matter a misleading emphasis. He writes:

> As I have often said, philosophy does not lead me to any renunciation, since I do not abstain from saying something, but rather abandon a certain combination of words as senseless. In another sense, however, philosophy requires a resignation, but one of feeling and not of intellect. And maybe that is what makes it so difficult for many. It can be difficult not to use an expression, just as it is difficult to hold back tears, or an outburst of rage.[2]

According to Wittgenstein then, philosophers are likely to feel an attachment to certain locutions. They have to fight this attachment and keep reminding themselves that they are nonsensical. But surely all that is necessary for them to encounter difficulty is that the locutions should still *seem meaningful*. There may be this attachment but there need not be. After all, the alleged nonsense might seem to a

2. From 'The Big Typescript', excerpted in *Ludwig Wittgenstein - Philosophical Occasions*, ed. James Klagge and Alfred Nordmann, Hackett, 1993, p.161.

philosopher to be false, perhaps horribly, harmfully, dangerously false. (Think of disputes between theists and atheists.) Far from being attached to it, he rejects it, though as false not meaningless. And would not most philosophers like to be rid of sceptical doubts? - it is their apparent meaningfulness and justifiability, not any emotional appeal, that causes the trouble.

The Absence Of Uncontroversial Examples Of Philosophical Nonsense.

I take it that everyone would admit that no philosophical utterances are *uncontroversially* nonsensical. But the point has both a trivial and a deeper interpretation. At the trivial level, the claim that something is philosophical nonsense involves the claim that someone has been taken in by an IOM, so at least one person must have failed to recognise something as nonsense. But almost everyone makes arithmetical mistakes, and yet there are uncontroversial examples of arithmetical mistakes, cases where everyone who has mastered arithmetic can be got to see that a mistake has been made (perhaps by himself). There is no parallel to this with philosophical nonsense and this is the deeper aspect of the point that there are no uncontroversial examples of it.

I doubt whether nonsensicalists themselves could agree on a single example.[3] A few decades ago it might have been possible to point to certain passages in Heidegger, but at least one Wittgensteinian now writes sympathetically of Heidegger, apparently finding his thought more congenial than that of an analytical philosopher like Davidson.[4] It might be replied that this is only what one would expect since being a nonsensicalist does not in itself commit one to any particular views about which utterances exemplify philosophical nonsense. And it is even possible that someone might accept the possibility of IOMs yet be unconvinced by *all* the accusations of talking nonsense that he has heard philosophers make (rather as someone might allow the possibility of unconscious motivation

3. One obvious and important disagreement is about whether philosophical nonsense is restricted to what is explicitly presented as philosophy. Some nonsensicalists, e.g. positivists, cheerfully attack religious discourse in the same way as they attack metaphysics; some Wittgensteinians, however, seem only to consider what philosophers say about such discourse as fair game, perhaps making anti-religious philosophers bear the brunt of their criticisms.

4. S. Mulhall, *On Being In The World: Wittgenstein and Heidegger On Seeing Aspects*, Routledge, 1990.

without being convinced by any of the actual examples alleged by Freud and others). This is true: but it hardly encourages optimism about the prospects for nonsensicalist methodology that there should be this lack of agreement.

Hacker and Baker mention a passage in which Wittgenstein suggests that one way of seeing the bogus character of philosophical problems is to tell oneself that if they had been genuine, they would have been solved long ago.[5] Suppose one were to reply that if they had been pseudo-problems, they would have been *dis*solved long ago (and to everyone's satisfaction). Wittgenstein could object that nonsensicalism is still relatively new and one should not be too impressed by its initial failure to secure agreement.[6] But there is something to be said on the other side too. If philosophical problems really are nonsensical, it ought to be possible to see this from an examination of these 'problems' *and nothing else*. But if they are genuine, one cannot say straight off what will turn out to be relevant to their solution. Consider two possibilities. First, some philosophical problems, such as the mind-body and freewill problems, seem entangled with empirical considerations, whereas others do not: but surely little is *obvious* here. (If you want an example of an area where it is controversial how important empirical issues are, consider the question: To what extent is an investigation of 'human nature' necessary in ethics?) Second, how well-founded is the assumption, so widespread today, that the investigation of language is crucial for philosophical progress? If nonsensicalism is mistaken, then one reason for thinking language important is mistaken, but there might be other, better reasons. Apart from a few uncontroversial points, such as that equivocation is a rich source of fallacious argument, not much is obvious here. What I am driving at is that if you think philosophical problems are genuine ones, it is not difficult to produce reasons why they have not yet been solved and why there is no agreement about to how to approach them.

In the last chapter I shall suggest that in general we just do not know how to deal with philosophical problems, how to solve or

5. *Analytical Commentary on the Philosophical Investigations*, Vol.I, Blackwell, 1980, p.486.

6. He makes a surprising concession in a comment recorded in Moore's report (Klagge and Nordmann, op.cit., p.113) - 'As regards his own work, he said that it did not matter whether his results were true or not: what mattered was that "a method had been found".' What justifies his confidence in a method which he admits may not yet have led to any correct conclusions?

*dis*solve them. The nonsensicalist can appeal to the novelty of nonsensicalism to explain away the failure to secure agreement about what is philosophical nonsense. But that is only one possible explanation: another is that alleged philosophical nonsense is not nonsense at all and philosophical problems are, though difficult, perfectly genuine.

Philosophical Nonsense Is Not Directly Exhibitable.

I mentioned arithmetical errors as errors of which there can be uncontroversial examples. It can be clear that someone has attempted to add up a column of figures and got the the wrong answer; and, assuming he has a basic competence in arithmetic, this can be made clear to him. It is noteworthy that we do not need to know *how* he came to make the error in order to know *that* he made it. If however we were educational psychologists interested in explaining arithmetical errors, we could investigate them secure in the knowledge that we could identify clear-cut examples.

Nothing like this will be possible with alleged philosophical nonsense. We could not be in a position to say that someone was clearly suffering from an IOM *however it might be explained.* We could not first identify the error and then try to explain it.[7] We might be able to say that what someone said was not a possible sentence of the language and it might be that we did not understand him. But, as I pointed out when introducing the Problem of Diagnosis, it would not follow that he did not mean anything, still less that he thought there was something he meant but was wrong. So we cannot say, 'IOMs clearly occur; let's try to explain them.' IOMs will always be something postulated. Not only are there no uncontroversial examples of philosophical nonsense, no utterance could *ever* show on its face its status as philosophical nonsense. So the nonsensicalist must first try to explain how IOMs are possible; if he succeeds, he can then perhaps use the possibility to shed light on philosophical perplexity.

---

7. Suppose a psychologist were to attempt to investigate empirically the conditions which encourage IOMs. The investigation would be likely to founder on the question: Who is to say what is nonsense? He might claim that under certain conditions his subjects wrongly saw meaning in certain locutions, but what if they refused to admit they were wrong?

What Do These Reflexions Suggest About Nonsensicalism?

In earlier chapters I have stressed how easy it is for nonsensicalists to slip into discussing what they officially claim is nonsense as if it made perfect sense and were at worst false. This fits comfortably with the staying power of alleged philosophical nonsense: it still seems to them to make sense and they have difficulty talking as if it did not. If we add to this the other points made in this chapter, that there are no uncontroversial examples of philosophical nonsense and that one could not identify examples of IOMs in advance of knowing how they are possible, it is natural to wonder why nonsensicalists are so confident in their nonsensicalism. Would not a possible explanation of their predicament be that nonsensicalism is mistaken: alleged philosophical nonsense does not just *seem* meaningful, it *is* meaningful, and the reason there are no *uncontroversial or obvious* examples of it is that there are no *genuine* examples of it?

Clearly, at this stage this is only a suggestion. As we have seen, the nonsensicalist can do something towards explaining away my observations. Indeed, in order to avoid giving the impression that I am prematurely concluding that it is all up with nonsensicalism, let me repeat something said in the first chapter. If the difficulties facing nonsensicalism are taken seriously and a real attempt is made to overcome them, it is possible that nonsensicalism will emerge strengthened. We might then be in a position to decide *which* nonsensicalist accusations are justified and a consensus might emerge.

Chapter Six: IOMs Outside Philosophy.

I f one is to be scrupulously fair to the nonsensicalist, one must consider non-philosophical cases where one might feel under some pressure to speak of 'illusions of meaning'. It might turn out that there is a better or clearer case for attributing IOMs to, say, schizophrenics than to philosophers. And it ought to be useful to compare alleged philosophical nonsense with nonsense, if such it be, encountered elsewhere: just how similar are they? Furthermore, one might worry that scepticism about the possibility of IOMs involves a kind of apriorism. Some very bizarre mental phenomena undoubtedly do occur[1] and one does not wish to ignore anything that looks like empirical evidence for IOMs.

a) Can One Have Nonsensical Dreams?

Malcolm writes:

> That something is implausible or impossible does not go to show that I did not dream it. In a dream I can do the impossible in every sense of the word. I can climb Everest without oxygen and I can square the circle. (Here there occurs the following footnote: What would be more senseless than to suppose that someone should not be able to distinguish propositions from tables? But Moore had a dream in which he could not do this.)[2]

Malcolm seems to be grouping empirical impossibility with more radical cases. Does he really think that the content of a dream can actually be senseless? After all, Moore might simply have dreamed that he was unable to *formulate* the difference between propositions and tables, and that doesn't seem senseless. It would have been a better example if he had dreamed that propositions *were* tables. But could anyone dream that? Someone might argue, 'If one says one dreamed that ... and follows this with a nonsensical string of words, one has *failed* to specify what one dreamed.' Someone else might allow that certain nonsensical combinations of words are possible specifications of the content of a dream but demand that they should at least look like meaningful sentences: 'I dreamed that ga-ga-ga' would not be an acceptable dream-report. Just what are the limits

---

1. See, for example, Oliver Sacks's The *Man Who Mistook His Wife For A Hat*, Duckworth, 1985. The title essay is as good an example of what I mean as any.
2. *Dreaming*, Routledge and Kegan Paul, 1959, p.57.

here? The answer, I take it, is not obvious. Since Malcolm appears to recognise the possibility of senseless dreams, let us begin by considering what his reasons might be.

He returns to the question of dreaming the impossible in a later chapter:

> I said previously that in a dream anything is possible. We can see why this is so. If we know that it is impossible for a certain thing to have occurred then the waking impression that it occurred is false, and we know therefore that one dreamt the impossible thing. Where the choice is between dream and reality the impossibility, in any sense, of a thing places it in a dream.[3]

I don't think this does explain why anything is possible in a dream. What it shows, as the final sentence makes clear, is that if one has already decided that something either occurred in waking life or was dreamed, then if one decides that it was impossible, one must assign it to a dream. But it does not show that there are no things 'too impossible' even for dreams. Still less does it tell us 'how much sense' a narrative must make to be acceptable as a dream-report.

Perhaps Malcolm is happy to accept impossible, even senseless, dreams because of his own idiosyncratic view of dreaming. He claims we should treat '*telling* a dream' as fundamental.[4] Now it is certainly possible to talk nonsense, if all that is meant is that what one actually says is not a possible sentence of the language. So, if someone awakes and talks nonsense, Malcolm might reason; 'He says it on waking. And being nonsense, *it can't be true*. So he is telling a dream.[5] This style of reasoning will not appeal to those with a more conventional view of dreaming.

Although I think that this may well be at the back of Malcolm's mind, I am not sure that it is the view he ought to take, given his view of dreaming. For is it really true that for him 'telling a dream' is fundamental? Is it not that, as the second passage quoted above suggests, sincere dream-telling is itself based on the waking

---

3. Ibid., p.65.

4. '[T]he concept of dreaming is derived, not from dreaming, but from descriptions of dreams, i.e. from the familiar phenomenon that we call "telling a dream".' (Ibid., p.55)

5. Malcolm (Ibid., p.85) places one restriction on what can count as a dream-report: telling a dream is telling a story. So, presumably, if he wants to allow for nonsensical dreams, he must say that the nonsensical waking report should in some way resemble a story.

impression that something happened? If so, then whether it is possible to dream that p, where 'p' is nonsensical, will depend on whether it is possible to have the impression that p, to seem to remember that p, where 'p' is nonsensical. And it is not obvious that it is or whether Malcolm would hold that it is.

Another possibility is this. Since Malcolm is a nonsensicalist, perhaps he holds that when people tell nonsensical dreams, they are suffering from IOMs. The dream does not really have a nonsensical content: rather, there is only the illusion of a content. This might seem the obvious line for him to take but I can find no textual evidence for it.

But, Malcolm aside, what should one say about 'nonsensical dreams'? There seem to be three possibilities:

(i) It just is possible to dream nonsense.[6] This meets the difficulty already mentioned: can one claim to have specified the content of a dream, if one's specification makes no sense?

(ii) Nonsensical dreams involve IOMs. Clearly one cannot assess this claim until one has decided about the possibility of IOMs. It would thus be inconclusive to cite apparently nonsensical dreams as empirical evidence for IOMs. There is another problem. Someone who claims to have dreamed that p, where 'p' seems nonsensical, is claiming to have had an *experience* (or at least to seem to remember having one). Could IOMs, even if they are possible, explain this? They are presented solely as *linguistic* illusions but in the present case more seems to be involved than seeming to see sense in a combination of words. If someone claimed to have dreamed that propositions were tables, say, could the view that he was mistaken in thinking he meant anything by 'Propositions are tables' do justice to this, even assuming it is correct?

(iii) If someone sincerely claims to have dreamed that p, where 'p' seems nonsense, there will always be a better description of the content of the dream (though it might not be easy to find) which does make sense. There is a suggestion of apriorism about this but it is perhaps what we think about other bizarre experiences. Take the common observation that 'the years seem to pass quicker as one gets

---

6. I suggest that if one does wish to defend this possibility, it should be called 'super-intentionality'. Take an ordinary intentional verb such as 'believe'. It does not follow from 'He believes that p' that 'p' is true (or that it is false). If 'dream' is a super-intentional verb, it does not follow from the meaningfulness and truth of 'He dreamed that p' that 'p' even makes sense, let alone is true.

older.' It is not in fact easy to replace this description with a non-paradoxical, logically impeccable, one. (Try it!) But perhaps we assume there will be one.

b) Can One Be Hoaxed Into Seeing Sense In Nonsense?

Consider the following episode: someone submits an article to a learned journal and has it accepted, but then triumphantly announces that it was constructed so as to be sheer nonsense and this shows the deplorably low standards prevailing in the discipline in question. What might have happened here?

He might have been using the word 'nonsense' in the colloquial sense to refer to gross falsehoods and inaccuracies where there is no doubt that something was meant. This seems to have been largely the case with the Sokal hoax.[7] Sokal filled what purported to be a post-structuralist critique of modern physics with elementary mathematical and scientific blunders and found that these escaped the notice of referees and editors. He mentions that he included a few meaningless sentences but, since he found he had no talent for inventing these, they formed only a small part of the whole.[8]

Let us assume that 'nonsense' was meant literally. Might the author have skilfully induced IOMs in his readers? The only other possibilities are (i) that they did not even *seem* to see meaning in the text but accepted it for some other reason and (ii) they really did see a meaning in it. Let us consider them in turn.

(i)I have stressed and shall be stressing again the difference between merely believing that a combination of words has meaning and believing that one means or understands something by it oneself. The present suggestion is that the editors merely had the first sort of belief. (I suppose it is possible that they saw through the deception and decided to play along with it but I shall ignore that.) People are quite easily led to believe that grammatical-looking sentences are meaningful, even though they have no idea *what* they mean: perhaps they think they use words in technical senses or extremely daring metaphors. This suggestion, however, implies a degree of irresponsibility on the

---

7. Discussed in *Intellectual Impostures*, Alan Sokal and Jean Bricmont, Profile Books, 1998.
8. Ibid., pp.248-49.

part of the editors that I am reluctant to postulate.[9] But I can imagine its being *part* of the explanation without this reflecting too severely on their intellectual integrity. Perhaps they thought they understood some of the text and assumed that the rest would yield to further study.[10] But then we must ask about the parts they thought they did understand: did they really see meaning there or did they suffer from IOMs? We must also ask about the text itself: was it supposed to be nonsense throughout or were chunks of nonsense embedded in a context of sense? If the latter, then perhaps they understood the parts that did make sense.

(ii) Perhaps the editors found a possible meaning in the text even though none had been intended. While this cannot be ruled out, it would seem to require great ingenuity on their part. Not only would they have to find meanings for the individual sentences but these meanings would have to cohere. (It would not do for the text to seem like a patchwork of possible but unrelated sentences.) I think we can say that this is unlikely.

So what then of IOMs as an explanation? Well, they will have to be possible if they are to help us with the case. But there is another difficulty that parallels the one just mentioned. Suppose IOMs are possible and suppose that someone reading a deliberately concocted farrago of nonsense experiences a succession of IOMs. What guarantee is there that the seeming (but really non-existent) meanings will appear to cohere, giving the impression of a unified text? None whatsoever, so far as I can see. So we still seem unable to explain the case satisfactorily. Or could one postulate a further illusion - of textual unity?

There is of course a possibility I have not considered: that hoaxes of the sort I have imagined just do not occur. Perhaps if one were to look closely at real cases, one would find that they were all of kinds I have been inclined to set aside: like the Sokal case or where the extreme irresponsibility I baulked at ascribing to the editors really is the explanation.

---

9. If instead of an article submitted to a learned journal, I had taken the case of a poem submitted to a literary journal, there would have been the possibility that the editors might plead that they thought the poem had aesthetic merit in spite of any difficulties there might be in saying precisely what it meant. This excuse is not available in respect of a text that has presumably been accepted for its intellectual and discursive content.

10. Perhaps it is worth mentioning here that wrestling with a recalcitrant text can sometimes stimulate one's own creativity, so that one emerges feeling that one has 'got something out of it'.

c) Do Schizophrenics Suffer From IOMs?

The utterances of schizophrenics often seem completely nonsensical – not just false, irrational or bizarre, but meaningless. Often grammar is disregarded and neologisms are used without explanation. An example of such a schizophrenic 'word-salad' begins:

> If things turn by rotation of agriculture or levels in regards and 'timed' to everything: I am re-fering to a previous document when I made some remarks that were facts also tested and there is another that concerns my daughter she has a *lobed* bottom right ear, her name being Mary Lou ... [11]

Here if anywhere one perhaps meets people who think they mean something when they mean nothing. Does the plain man view schizophrenics thus? I am not sure. Perhaps he is content to say that he does not understand them and leaves it to the expert to decide whether there is any meaning there.

There have certainly been attempts to show that schizophrenic utterances can be interpreted if one makes the effort. Perhaps the best known is that of R.D. Laing. This has come to be associated with a bewildering array of other doctrines espoused at various times by him and his co-workers—that schizophrenia is a condition of the family and not of the individual; that it is not something that befalls one but a way of being; that it is not an illness; that it is a comprehensible reaction to an intolerable situation; that a schizophrenic episode can be a healing experience; that schizophrenics are often less estranged from reality than many judged sane—and perhaps also the view of Thomas Szasz that there is no such thing as mental illness.[12]

I think it important to keep these claims apart. In particular, there seems no inconsistency in maintaining *both* that schizophrenia is an illness affecting the individual *and* that what schizophrenics say is intelligible. This indeed seems to be Laing's own position in his early book, *The Divided Self*.[13] Imagine that it were conclusively proved that schizophrenia is caused by a chemical imbalance in the brain and that this in turn is an inherited condition. That would seem to be a pretty thoroughgoing defeat

11. Quoted in Jennifer Radden, *Madness and Reason*, George, Allen and Unwin, 1985, p.75.
12. See his *Myth of Mental Illness*, Secker and Warburg, 1962.
13. Pelican, 1965

for the Laingian approach taken as a whole; yet he could still be right that what schizophrenics say can be understood.

If there is a problem in explaining how someone can be mistaken in thinking he means anything, there will be a problem even if that person is schizophrenic. It is possible that the schizophrenic produces a stream of words without even thinking he means anything: perhaps he finds his vocal organs working against his will. But I know of no reason to think that this is always, or even usually, the case and there are plenty of autobiographical accounts of schizophrenic episodes. I don't wish to dogmatise. There are those cases that would once have been regarded as instances of demonic possession.[14] Perhaps in these the victim does not necessarily think he understands what comes from his own mouth. But, so far as I know, schizophrenics do not normally think of themselves as talking nonsense at the time.[15] So in most cases we seem faced with the alternatives: either they mean something by what they say (though they may be using language non-standardly) or they are suffering from IOMs.

There are a number of problems and unclarities about the proposal that we try to understand schizophrenic speech that need to be mentioned:

(i) I suspect that Laing and his followers exaggerate the extent to which psychiatrists of other persuasions have dismissed schizophrenic utterances as sheer nonsense. After all, classifying patients by the nature of their delusions, as is often done, presupposes that one finds some measure of intelligibility in what they say. And certainly theorists in the Freudian, Jungian and Kleinian traditions have attempted to interpret schizophrenia and schizophrenic utterances. One can no doubt find cases where psychiatrists seem to dismiss these utterances as unworthy of serious consideration, but I suspect that this is more a matter of intellectual laziness than of theoretical conviction. Or perhaps some busy psychiatrists would argue that if they can control a schizophrenic's symptoms by means of drugs, this relieves them of the obligation to

14. 'A most curious phenomenon of the personality, one which has been observed for centuries, but which has not received its full explanation, is that in which the individual seems to be the vehicle of a personality that is not his own.' (Ibid., p.58)

15. It could perhaps be argued that if someone's behaviour and utterances are sufficiently bizarre, one cannot confidently attribute to him a belief (true or false) that he means anything by what he says. I do not know whether anyone has ever taken this view. One thing is clear: it would not be to attribute IOMs to the sufferer.

try to understand his utterances when not on medication.

(ii) If one tries to understand schizophrenic utterances by attributing *unconscious* meanings to them, as Freudians might, one is likely to arrive at an understanding of them that is not the schizophrenic's own understanding (assuming there is one): one has bypassed the question, 'What if anything does the patient mean?', taken in the sense with which we are concerned. So far as I can see, Laing's own approach does not run into this difficulty.

(iii) It does not follow that if one could understand what the schizophrenic says, one would thereby understand his condition (nor vice versa). Schizophrenia might involve some impairment in the capacity to communicate and what the schizophrenic is trying to say might often be something quite ordinary, which sheds no light on his condition. Yet those who try to understand the utterances always seem to assume that this is closely bound up with understanding the nature and genesis of the condition. Laing is a case in point. But the assumption is unjustified. The delirium characteristic of high fever often takes the form of utterances not dissimilar to those of schizophrenics, yet no one would assume that understanding them would help one understand the aetiology of the fever (though it might tell one *something* about the patient).

But in spite of any caveats and reservations, my sympathies lie with those who have tried to understand schizophrenic speech. Intelligibility is unlikely to be found if no one looks for it. Would the Egyptian hieroglyphs have been deciphered if they had been dismissed as merely decorative? The difficulties should not be underestimated, however, as the specimen of a schizophrenic 'word-salad' with which I opened this section makes clear. But one could not, faced with this apparent unintelligibility, simply regard it as evidence for the existence of IOMs: rather, one needs to know whether IOMs are possible before one can postulate them as an explanation of it.

d) Can One Have IOMs Under The Influence of Drugs?

Sometimes people under the influence of drugs have thoughts which seem to them profoundly significant, perhaps the 'Secret of the Universe'. They hasten to record them and later find that what they have written seems sheer nonsense. E.M. Thornton gives some examples culled from William James and elsewhere - 'What's a mistake but a kind of take?'; 'What's nausea but a kind of -ausea?'; 'Emphasis, EMphasis, there must be some emphasis in order for there to be a phasis'; 'By George, nothing but othing'; 'That sounds like nonsense, but it's pure *on*sense'; 'The banana is great, but the skin is greater'.[16] One wonders whether it might be possible to produce this kind of experience without the use of drugs, by hypnosis for example, telling the subject that he will see a profound meaning in some piece of gibberish or even in the stains on the ceiling.

The striking fact about this sort of case is that usually the recipient of the revelation, once he has come round, no longer even *seems* to see meaning in the locution in question. This sets them apart from alleged instances of philosophical nonsense, a point to which I shall return. But could it also provide evidence that we are dealing with *illusions* of meaning rather than cases in which the person means something but expresses himself badly?

Now our discussion of the possibility of IOMs is a conceptual investigation. So we cannot simply adduce empirical evidence for their occurrence. Nevertheless, if I had to say what was the nearest thing I knew to empirical evidence, it would be these utterances – not those of Hegel, Heidegger, Derrida or theologians holding forth on the Trinity. Philosophers may have originated the notion of IOMs but it may not be in philosophy that the most plausible cases are found. It might even turn out that they are possible but do not occur in philosophy.

One can avoid accepting these drug-induced utterances as *prima facie* evidence for IOMs by assuming that the drug-taker gives a non-standard meaning to certain words in the problematic locution and then *forgets what he meant* when he comes round. This would explain his having found it meaningful in the sense that he meant *something*, though not necessarily his having found it tremendously significant. But it is only the former that we are trying to explain here. It is no doubt possible for something to have a completely illusory aura of importance but that, if it is a problem, is a different one.[17]

Nevertheless, the suggestion of apriorism that we have encountered

---

16. *The Freudian Fallacy: Freud and Cocaine*, Paladin, 1986, pp.168-70.

17. It is noteworthy that among examples of drug-induced 'nonsense' locutions that seem banal are found alongside ones that seem strictly meaningless. Some are difficult to place. Does 'The banana is great, but the skin is greater' seem meaningless, banal, or perhaps not quite either? To a Wittgensteinian, of course, banalities will be a species of nonsense if they are uttered outside a suitable context.

already is rather marked here. Who really knows what is going on in these bizarre cases? I prefer to leave it open whether they involve IOMs. What matters is whether the conceptual difficulties can be overcome. If they can, then probably we have here evidence for the actual occurrence of IOMs; if they cannot, then some other explanation of these cases, however contrived and aprioristic it might seem, will have to be found. I believe that nonsensicalists are themselves guilty of a certain apriorism: they assume the possibility of IOMs more because they feel a philosophical need for them than because they have done much to confront the problems with the notion. I do not wish to seem equally prejudiced, but *against* nonsensicalism.

What Are We To Conclude From These Cases?

I have discussed four kinds of non-philosophical case where one might suspect that IOMs occur. There are others. Wittgenstein mentions grammatical jokes (*PI* 111). Could it be that a grammatical joke involves a momentary IOM which is dispelled when one *sees* the joke and could it be that when a grammatical joke becomes stale is because one no longer even momentarily seems to see meaning in the risible locution?

I think however that the cases I have discussed are sufficient for my purposes. One conclusion to be drawn from them, which might have been expected from what was said in the last chapter, is that one cannot argue: IOMs must be possible because actual examples exist. Rather one must decide whether they are possible before one can know whether to trust any inclination one might feel to see these cases as exemplifying them. This is perhaps so obvious that it might be wondered why I have gone into detail about the cases. There are two reasons.

First, their intrinsic interest: it is surely important that the idea of IOMs, if sound, will not be just a technicality of philosophy but one that might have application elsewhere.

Second, there is a striking contrast between these cases and ones of alleged philosophical nonsense. It might be summed up in the thesis: philosophical nonsense must surely be a far better imitation of sense than the other kinds.

I have pointed out the staying power of philosophical nonsense. This contrasts with drug-induced cases where typically the one who seemed to see a deep meaning in what to others is gibberish no longer does so when the effects of the drug have worn off. It is not easy to say whether the other cases of non-philosophical nonsense contrast so sharply with the philosophical, largely, I think, because the relevant research has not been done or is not well-known. For example, have schizophrenics who have been cured or who are in remission ever been confronted with the seeming gibberish they produced at the height of their illness and asked to interpret it?

Another feature of philosophical nonsense is its apparent communicability. People can easily be induced to see sense in scepticism and in many metaphysical doctrines. There are of course notoriously obscure philosophers but there are others with a reputation for clarity and these too have not escaped charges of talking nonsense.[18] But the utterances of schizophrenics and drug-takers often convey nothing to others and probably only certain people are vulnerable to hoaxing with nonsense-texts. (I am not sure what to say of dreams.) It is worth adding that one could put this point about the communicability of philosophical nonsense even more strongly by saying that the IOMs in question are *shared*: recall that people often formulate what seem to be essentially the same philosophical theses, questions and doubts independently.

Finally, philosophical discourse seems to consist of reasoning. It is or appears to be an organised body of inferentially linked assertions and questions. Non-philosophical nonsense is not usually like this, though of course a hoax-text might mimic a philosophical one.

It is clear that where one can point to a contrast between philosophical and non-philosophical nonsense, the former comes off better. Perhaps this was only to be expected. Philosophical nonsense has to be something capable of taking in intelligent people in full possession of their faculties. If it is an aberration, it is not a *mere* aberration: it is, for example, better to be a metaphysician than a madman. But one might wonder whether alleged philosophical nonsense seems to be such a good imitation of sense because it is not nonsense at all. And one might also ask: even if IOMs are possible, is philosophy the place to look for them?

One final observation: no clear answer has emerged to the question whether non-philosophers recognise the possibility of IOMs. What is

---

18. For many years the really obscure philosophers tended to be simply ignored by philosophers in the Anglo-Saxon tradition, so that (paradoxically, perhaps, or ironically) there may have been more accusations of talking nonsense made against Descartes or the British Empiricists than against the likes of Hegel and Heidegger.

clear is that in the 'language-game' with the word 'mean', although they say 'This does not mean anything' and 'He did not mean anything by what he said', they do not often say, 'He did not mean anything by what he said, *even though he thought he did.*' Whether they *ever* say it or anything like it - of schizophrenics and. drug-takers, for example - remains unclear.

Chapter Seven: How, If At All, Might IOMs Be Possible?

When someone experiences a perceptual illusion he is normally able to say what it is he thinks he sees, hears, feels, etc. There are no doubt cases where someone has difficulty describing what he seems to perceive but then there are cases of veridical perception of which this is true. There is nothing about the fact that a perceptual experience is unveridical that makes it especially difficult for the victim to say what he seems to perceive - or for others to report it.

Contrast this with IOMs. If one claims that someone is mistaken in thinking he means *anything* by what he says, one cannot go on to say *what he thinks he means*. That would surely be to say what he does mean. At least I cannot imagine why anyone should want to credit someone with thinking he meant such-and-such (something that can be specified) and yet deny that he had actually succeeded in meaning that.[1] But even if I am wrong in this, it is clear that with philosophical accusations of talking nonsense, if one could say what the accused *thought he meant*, the accusation would have failed, for one would have elicited a formulable and discussable thesis, one that was not nonsense.

There is thus a sense in which IOMs would be illusions without content. Even if there is some way of saying what is going on in the victim's mind, it will not be a straight statement of what he means (or thinks he means). This connects with the peculiarly annihilating effect that nonsensicalist accusations would, if justified, have. In fact, 'annihilating' is not quite the word, for it implies that one destroys, makes *something* into *nothing*, whereas a justified nonsensicalist accusation would show that all one ever had was a nothing; it seemed there was something there, something meant, something that could be discussed, but there wasn't. Anscombe puts it thus:

> This idea of philosophic truth [that in the *Tractatus*] would explain one feature of philosophy: what a philosopher declares to be philosophically false is supposed not to be

---

1. If one wishes to pursue this question, one might like to consider *PI* 540-42 where Wittgenstein seems to be saying that someone can wrongly think there is something he means by an utterance and *at a slightly later time* say what it was that he now thinks he meant.

possible or even really conceivable; the false ideas which he conceives himself to be attacking must be presented as chimaeras, as not really thinkable thoughts at all. Or, as Wittgenstein put it: An *impossible* thought is an impossible *thought* (5.61) - and that is why it is not possible to say what it is that cannot be thought; it can only be forms of words or suggestions of the imagination that are attacked. Aristotle rejecting separate forms, Hume rejecting substance, exemplify the difficulty: if you want to argue that something is a philosophical illusion, you cannot treat it as a false hypothesis. Even if for purposes of argument you bring it into contempt by treating it as an hypothesis, what you infer from it is not a contradiction but an incoherence.[2]

The victim of a philosophical illusion must have been wrong to think there was anything he meant by his question or thesis. His supposed error seems the most radical it is possible to make, worse than asserting the most egregious falsehood, even a *necessary* falsehood. (Anscombe begins by talking of the 'philosophically false' but goes on to argue that it is not *falsehood* that is in question.)

I have been trying gradually to bring out just how astonishing nonsensicalist accusations are. That IOMs must be illusions without content is perhaps the best illustration of this so far (especially if one contrasts it with the 'staying power' of philosophical nonsense). It is natural to wonder whether the point could be used to refute nonsensicalism, to demonstrate that there just could not be IOMs. Such a knock-out blow is not, I think, possible and it is instructive to see why.

It is certainly difficult to get one's mind round the suggestion that, though one thinks one means something, one means nothing. One is

2. *Introduction To Wittgenstein's Tractatus*, Hutchinson, 1959, p.163. John W. Cook in 'Wittgenstein on Privacy', *Philosophical Review*, Vol. LXXIV (1965), p.281, after noting that some philosophers have complained that Wittgenstein attacks the notion of 'private language' without saying clearly what he means by the term, replies that Wittgenstein 'does not try to make this clear because the idea under investigation turns out to be irremediably confused and hence can only be suggested, not clearly explained.' And Crispin Wright in *Wittgenstein on the Foundations of Mathematics*, Duckworth, 1980, p.205, remarks that 'things have come to a pretty pass if a philosopher cannot legitimately reject a concept without first supplying a full explanation of it - his rejection, after all, may be based on the belief that a satisfactory explanation cannot be given.' Obviously passages such as these raise many of the questions with which I am concerned throughout the book. Here I use them as graphic illustrations of the point that IOMs must be conceived of as illusions without content.

inclined to object, 'No matter how badly I may have expressed myself, I must mean *something*, if I think I do'; 'There just seems *no room* for error here'[3]; 'I can no more think I mean something when I mean nothing than I can think I am thinking something when I am thinking nothing'. But this last protest reveals a difficulty that the defender of IOMs can exploit. He can reply:

> I agree that *thought* must always have a content but that does not show that there cannot be IOMs. For it might still be possible for someone to be wrong in thinking he means anything by an utterance, if he is deceived by his having some other thought. IOMs are illusions without content in that one cannot say what the victim of one thinks he means. But that is not to say that things going through his mind which do 'have content' do not lead him to think mistakenly that there is something he means. As regards your other protests, they amount to 'I don't see *how* one could be wrong' and that is not a proof that there is no way.

I can see no obvious counter to this. But at least the onus is now clearly on the nonsensicalist to tell us what these misleading mental occurrences are and how they mislead. All too often nonsensicalists write as if the onus were on those accused of talking nonsense to explain what they mean in accordance with conditions imposed by the accuser.[4]

I am not in fact aware that anyone has ever tried to show that IOMs are impossible because they would need to be illusions without content, though I could imagine a Phenomenologist arguing thus. Another way of bringing out the difficulty facing such attempts is by considering Descartes' universal doubt. Kenny chides Descartes for

---

3. At *PI* 288 Wittgenstein suggests that we would be baffled if someone claimed to know the meaning of the word 'pain' but not to know whether he himself were there and then in pain. But I know of nowhere where he acknowledges that there is also something odd in the idea of someone's being in doubt about whether he himself means anything by what he is now saying. It is no answer to point out that it would be self-defeating to accuse a person entertaining such a doubt of being a victim of an IOM. There is this danger, certainly, but it is still important not just to *ignore* the oddity of the doubt. We must ask what kind of uncertainty, if any, is possible about whether one means anything oneself.

4. I am prepared to admit that this bias may have had some beneficial effects. It has no doubt helped to keep Anglo-Saxon philosophy out of the hands of those who appear to equate obscurity with profundity. But I would still insist that if one accuses someone - even a Heidegger - of being a victim of IOMs, one must explain how these are possible.

not doubting that he knows the meanings of the words he uses and argues that if he had done so, he would have found himself unable to formulate his universal doubt.[5] Suppose Descartes had thought of this and replied that, so long as he knew what he meant by his words, it would not matter whether he were using them correctly, in accordance with the rules of French or Latin, when formulating his doubts *to himself*. He might have said, 'Even an all-powerful demon could not deceive me into thinking I meant something by a locution when I meant nothing' (just as he denied that the demon could deceive him into thinking he was thinking when he was not). But could he have ruled out the possibility that the demon produce in him *thoughts of some sort* which misled him into thinking there was something he meant? To me at least the answer is not obvious.

What Has Been Done Towards Explaining How IOMs Might Be Possible?

I shall consider those suggestions I know of but I have to say that I find them curiously undeveloped and unsystematically presented. This is puzzling. One would have thought that the importance of IOMs to those philosophers who regularly accuse others of talking nonsense was so great that the issue would have been confronted head-on. It surely isn't *obvious* that there can be IOMs.

Anyone approaching the problem from the direction of other 'How is ... possible?' questions in the philosophy of mind should be struck by how different is the treatment of IOMs. Even the most resolute defenders of the possibility of weakness of will, say, or self-deception concede that sceptics are right to be puzzled. There is a consensus that here are problems crying out for resolution. There are obvious *prima facie* difficulties in explaining how someone could act against his own present better judgment or do anything that could appropriately be called 'deceiving himself'. Yet defenders of these possibilities have an advantage that defenders of IOMs must lack: they can claim to be defending common sense. In daily life we do seem to allow these possibilities.[6] But, as I have pointed out, it is

---

5. *Wittgenstein*, Penguin, 1975, p.206.

6. In the first chapter I noted that the idea of *unconscious motivation* has been greeted more critically than has that of IOMs. It is not part of common sense in quite the way that the ideas of self-deception and weakness of will are, but one could say that it has been received fairly easily into the educated person's intellectual armoury (perhaps because it seems to be linked to the idea of self-deception).

not clear that outside philosophy there is any general belief in IOMs. It is curious therefore that the novel idea of IOMs has been received in a less critical spirit than the familiar ideas of self-deception and weakness of will.

I shall not embark on a detailed critique of the various suggested ways of explaining IOMs. My reasons will only fully emerge in the course of this chapter and the next, but briefly they are the following:

(i) I am not sure that I have uncovered all the suggestions that have been made.

(ii) I am not always sure that I am correct in finding a given suggestion in the texts I discuss.

(iii) Many of the suggestions I have been unable to take very far or to state precisely enough for there to be no doubt that they bear upon our problem.

(iv) If I were to attempt to evaluate them fully, I would have to consider the possibility of *combining* them and my task would become immense.

(v) In the next chapter I consider what I believe is a more Wittgensteinian approach to the problem than any of the individual suggestions considered in this. (In a sense, perhaps, it combines them or some of them.)

The problem then is this: can we make clear how someone could think there was something he meant when in fact there was nothing he meant? Just what could be going on in such a case? Wittgenstein seems to object to questions of the form, 'What goes on when ...?', asked about the psychological, a point to which I shall return in the next chapter. For the present let me put the question in another way. If IOMs are possible, it ought sometimes to be correct to say, 'I used to think I meant something by '*p*'; I now see I only ...' Only what?

Goebbels is supposed to have said that if you tell people something often enough, they will come to believe it (even if it is false). Suppose something similar were said of nonsense: repeat a piece of nonsense often enough and people will come to seem to see a sense in it. Such a claim would shed no light on our problem for it *assumes* that IOMs are possible. *If* they are, *then* it could be a true claim. This may sound obvious; but I have known it be claimed as an explanation of how IOMs might occur that if one already wrongly thinks a locution *has meaning*, then one will be more likely to seem to see a meaning in it oneself. Again, this could be a true empirical claim - but only if IOMs are possible. This is just to say that we are dealing with a *philosophical* problem, not one that can safely be left to the psychologists.

Suggested Explanations.
I. An Account In Terms Of Emotions.

It is not uncommon for philosophers to remark that, although a locution might evoke in one various associations, emotions and images, this does not show that one means or understands anything by it. For example, Carnap[7] argues that images and feelings '[f]rom an earlier period of significant use' can remain deceptively associated with a word after it has ceased to be meaningful; P.T. Geach[8] maintains that '[w]e can take a man's word for it that a linguistic expression has given him some private experience - e.g. has revived a painful memory, evoked a visual image, or given him a thrill ... [but not] that he attached a sense to the expression, even if we accept his *bona fides* ...'; and Wittgenstein (*BB*, p.65) distinguishes an expression's producing certain experiences from its actually having a use. In this section and the next I consider the suggestion that it is *because* locutions evoke emotions and images that one can be mistaken in thinking one means anything by them.

Consider the positivist claim that unverifiable utterances are *factually* (cognitively, empirically) meaningless. Put like that, the claim seems to be that they might have one sort of meaning but lack another, judged more important (and no doubt that they masquerade as having the higher grade of meaning). But this is not always clearly distinguished from the claim that an utterance can be *completely* meaningless but seem meaningful because one is duped by some associated mental phenomenon. Thus the claim that it has 'emotive' but not 'factual' meaning is not always clearly distinguished from the claim some emotional perturbation felt in connexion with it misleads one into thinking there is something one means by it. The latter claim would amount to a suggestion about how IOMs are possible. For example, it might be held that a person is misled into thinking there is something he means by 'God created the Universe' by a feeling associated with that locution. If one is going to assign this deceptive role to emotions, one must be careful with the objects of the emotions one appeals to. One could not, surely, explain

---

7. 'The Elimination of Metaphysics through Logical Analysis of Language', reprinted in *Logical Positivism*, ed. A.J. Ayer, Greenwood Press, 1959, p66.

8. *God and the Soul*, Routledge and Kegan Paul, 1969, p.20.

someone's wrongly thinking he meant something by '*p*' by appealing to an emotion that had as its propositional object: *p*. But one would not be debarred from appealing to an emotion with some other object. It is not at all clear, however, why or how emotions could be capable of playing this deceptive role. There is a danger that the explanation will be as mysterious as the phenomenon it is supposed to explain.

II. <u>An Account In Terms Of Mental Images.</u>

Wittgenstein often suggests that we are misled by mental pictures - sometimes he seems to imply that such pictures could explain IOMs. Passages dealing with the pernicious influence of pictures are common: he suggests that sometimes we advert to pictures to *assure* ourselves that we mean something (*Z* 251); that misleading pictures are embedded in language (*PI* 115, 295); that some pictures seem to suggest how an expression might be used without really doing so (*PI* 352). But the clearest statement I know that pictures can mislead us into thinking there is something we mean when there is not is in 'Notes for Lectures on "Private Experience" and "Sense Data" '[9]—' "It seems to me to have sense." You are undoubtedly using a picture, therefore it "seems to you to have sense".' I shall in section VI ask whether there might be other ways of taking Wittgenstein's references to 'pictures', but it is surely natural to try taking them in the first instance as suggestions that it is mental *images* that mislead.

Two questions about the formulation of this suggestion must be asked. Is it being claimed that people mistake mental images for 'meanings'? Or would that be too reificatory? Would it suggest that meanings are introspectible entities that can be confused with mental images (an assumption that is dubious and would certainly be rejected by Wittgenstein)? It is perhaps safer to say: the victim of an IOM confuses *the fact that* he associates a mental image with a locution with his meaning something by it. (It might perhaps be added that in some cases the victim of an IOM holds a theory about meaning that encourages the illusion: he might, for example, think that meanings *just are* mental images.)

The other question is about *when* images mislead. It is not

_____

9. Reprinted in *Philosophical Occasions*, ed. James Klagge and Alfred Nordmann, Hackett, 1993, p.238.

plausible to claim that if someone associates a mental image with a meaningless locution, he must of necessity experience an IOM. Would just any image do? Surely not. One would have to say how the image must be related to the locution and do so without surreptitiously assigning a meaning to the locution, making the image the victim of the IOM associates with it depend on what he *understands* by it. A closely related point is that one must remember *what* one is trying to explain. After reading *PI* 352 one might offer the following account:

> Someone thinks he means something by '7777 occurs somewhere in the expansion of π' because he imagines a series of which one person sees the whole but someone else sees only a part, and vaguely supposes that some godlike being might see the whole of an infinite series of which we see only a part.

This may explain the illusion of *verification* but hardly the illusion of seeing meaning in the unverifiable claim.

III. Misleading Grammatical Analogies; Language-Game Conflations; Category Mistakes.

It is widely held that philosophers are misled into treating the meaningless as meaningful by deceptive grammatical analogies. They treat a word as if it had the grammar of another whose grammar is in fact only superficially similar.[10] Now I have to say at the outset that I have a problem about the interpretation of this suggestion and that a similar problem arises with the next two. Suppose it were said, for example, that if A (x), B (x) and C (x) make sense and so do A(y) and B(y) then we are likely to assume that C(y) makes sense, even if it does not. What is the mistake we are supposed to be making? Is it that we will expect C(y) to *have a meaning*? Or are we supposed to seem to see a meaning in it ourselves, to think that we ourselves mean or understand something by it? I can see that the first error might sometimes occur, though whether it will be common in philosophy is less obvious. Perhaps

---

10. There are various forms which this idea takes, depending upon whether it is expressed in terms of 'language-games', 'grammar' or 'categories'. The basic idea is that a locution can be constructed to look and sound like - have the *Satzklang* of - a genuine sentence without really being one. I have decided not to burden the text with a mass of references since, as I shall explain, I am not sure how far the idea (which is in any case a familiar one) is intended as an answer to our question.

philosophers who are committed to particular theories of meaning might believe that all locutions constructed after a particular pattern ought to be meaningful and this might involve them in mistakenly expecting a certain locution to be meaningful. (As pointed out in Chapter Four, it is easy to think of non-philosophical examples of this error: one might be misled into thinking a locution a sentence of a foreign language by its resemblance to genuine sentences of that language.)

But could the suggestion explain the second error? How is a grammatical resemblance to a meaningful locution, however close, supposed to make one think one means or understands something by a meaningless one? There is some degree of analogy between 'Quadruplicity drinks procrastination' and 'John drinks beer', but most people do not even *seem to see* meaning in the former. Perhaps the analogy is of the wrong type or insufficiently close but this only emphasises the problem: if we have a case in which analogies are close enough to mislead, how does the deception work? Many, perhaps most, philosophers think they mean something by, 'It is possible that I am now dreaming. How can I tell whether I am or not?' According to Malcolm, they are wrong. If he were to claim that they are misled by a resemblance to such meaningful locutions as 'It is possible that I am now hallucinating, etc.', he would have to say *how*. Simply noting the resemblance of a meaningless locution to a meaningful one is not to think one understands it, nor is constructing a meaningless locution on analogy with a meaningful one to think one means something by it.[11]

Something seems to be missing from the suggestion if it really is supposed to explain IOMs. But perhaps the nonsensicalist will reply that it is not intended to stand alone. Perhaps he will admit that misleading grammatical analogies can at most explain how someone might come to believe that a meaningless locution must have a meaning and that some other factor will have to come into play if the dupe goes on to think that he means something by it himself. One of the other suggestions, that involving mental images for example, might have to be appealed to.

IV. 'An Atmosphere Accompanying The Word.'

At *PI* 117 Wittgenstein writes:

> You say to me: 'You understand this expression, don't you?

---

11. The point seems to me to apply irrespective of whether it is held that one is *aware* of being influenced by analogies with meaningful locutions (a question nonsensicalists do not seem to have much to say about).

> Well then—I am using it in the sense you are familiar
> with.'—As if the sense were an atmosphere accompanying
> the word, which it carried with it into every kind of
> application.

This could be taken as a suggestion about how someone could
think there was something he meant by a locution when there was
not. He transfers a word from a context in which its use is familiar to
one in which it has no established use but fails to realise this. He
thinks he is using it in the old way but is not really using it in any
clear way at all. This suggestion could be taken as a variant of the
last, for the transfer of a word to a new and problematic context
might well be thought more likely if the new context looks
grammatically analogous to the old. However, the
atmosphere-metaphor introduces a new element: philosophers are apt
to assume that a word, if it has a meaning in one context, will
without further ado carry its meaning into new contexts.

But the same problem arises as with the last suggestion.
Philosophers might expect the results of such transferences to be
meaningful; but this hardly explains their thinking that they
themselves mean something by them, if in fact they do not. The
suggestion needs to be supplemented with an account of what
happens when a philosopher comes himself to seem to see
meaning in a locution which the error in question has led him to
expect to have meaning.

V. The Context Of The Utterance.

The passage just quoted, *PI* 117, continues:

> If, for example, someone says that the sentence 'This is
> here' (saying which he points to an object in front of him)
> makes sense to him, then he should ask himself in what
> special circumstances this sentence is actually used. There it
> does make sense.

Although this reads as though Wittgenstein thought it an expansion
of what has gone before, it seems to me to involve a change of
subject. He has been talking about the context of *a word or
expression* within a sentence, whereas now he is talking about the
context, i.e. situation, in which *a sentence* is used. This shift from
verbal to situational context might suggest a quite different account
of how IOMs could occur. The idea would be that philosophers take
whole sentences that have non-philosophical uses, detach them from

the contexts in which they have these uses and, without specifying any new uses, think they still mean something.[12] But we meet the same difficulty as before. It could be put thus: does the 'they' at the end of the last paragraph refer to the philosophers or the sentences? Perhaps philosophers do treat sentences in this way and perhaps this is a mistake (though neither claim is *obviously* true); but could this on its own explain their thinking that they themselves mean something by the sentences detached from their normal contexts, as distinct from merely expecting them to have meaning? If the fact of the matter is that they do not mean anything, have we been given any real explanation of why they think they do?

VI. Metaphor and Pseudo-Metaphor.

There may be another way of taking Wittgenstein's references to the misleading influence of pictures. Perhaps he is concerned with the misleading influence of *metaphor*. Although metaphors normally do conjure up sensory, especially visual, images in the mind of one who appreciates them, it is worth asking whether there might be ways in which metaphors could produce IOMs which do not involve this power.

One idea that is surely present in the later Wittgenstein is that there are metaphors embedded in ordinary language which can mislead if taken literally. Could they produce IOMs? Consider the spatial metaphors that are common in our talk of the mental. Perhaps someone can talk nonsense if he takes a phrase like 'at the back of one's mind' literally.[13] But there is an obvious difficulty: if it really is possible to take a phrase literally, then utterances based on this will not be strictly nonsense, though they may be false. Compare 'My heart is aching' with 'Life is a bowl of cherries'. Both are conventional metaphors but the former might conceivably be misunderstood and taken literally to refer to a medical problem, whereas it is not clear that there is *any* literal way of taking the latter. So what should one say of

---

12. This idea is especially prominent in *On Certainty* (e.g. 348, 350, 464, 622). I get the impression that it took on increasing importance in his later years, that he became less insistent that the locutions he rejected as nonsense have no use in the language and more inclined to say that it is when philosophers utter them that they are deprived of meaning.

13. See J.F.M. Hunter, *Wittgenstein on Words as Instruments*, Edinburgh U.P., 1990, pp.69-86.

someone misled by 'at the back of one's mind'? That he *tries* to take it literally? That he *believes* he is taking it literally? I am not sure but, whatever one says, one must - if the suggestion is to help explain IOMs - avoid implying that there is a genuine literal interpretation available.[14] But, assuming one succeeds in this, there is a further difficulty. The victim of the IOM must mistakenly think he sees a literal interpretation of such a sentence as 'The idea was at the back of my all the time' in addition to or instead of the ordinary interpretation of it. But is this not just an *instance* of our problem rather than a solution to it? It is not clear that we have made any progress towards explaining how he can mistakenly think he sees a possible literal sense in it if there just isn't one.

Perhaps there is a better way of using the notion of metaphor to vindicate the possibility of IOMs. Could it be that sometimes people talk nonsense because they think they are saying something metaphorically that could if necessary be said literally, when in fact there is no way of saying 'it' literally? They think they are speaking of X by speaking of Y which resembles it in certain ways when, if they were to try to specify these ways by describing X and Y separately, they would be unable be to do so. Certainly it is often difficult to purge what one says of metaphor: if someone keeps rejecting one's formulations as *merely* metaphorical and demands a literal statement, one can quickly come to feel as Socrates' interlocutors must have done when continually pressed for definitions. But this in itself raises a doubt. Can everything that can be said be said literally? Perhaps there are things that can only be said metaphorically: in view of the pervasiveness of metaphors drawn from the physical world in descriptions of the psychological, some have felt that talk of the psychological is irreducibly metaphorical.[15] If we were to reject as nonsense everything that we could not express literally, we might find ourselves rejecting more than we had bargained for.

Nevertheless, I am prepared to let this metaphor (really: pseudo-metaphor) account stand as a contribution to our problem. It is highly programmatic, but so are many of the other suggestions in this chapter. Perhaps a way will be found of purging psychological

---

14. Cook (op.cit.) is especially clear about the need to avoid this error, though he discusses a rather different kind of case. It is another example of the 'sense that is senseless' trap.

15. See P.T. Geach, *Mental Acts*, Routledge and Kegan Paul, 1957.

language of metaphor[16], and if no way is found of doing this with, say, theological language, there might be a case for calling the latter 'nonsense'. No one, I dare say, is entirely happy with the irreducibly metaphorical. At any rate, the matter is still sufficiently obscure to discourage hasty decisions.

VII. General Misconceptions About Language.

It seems likely that Wittgenstein thought that mistaken views about how words have meaning can lead one to talk nonsense. There do in fact seem to be connexions between some of the suggestions about how IOMs might be possible and certain views about what meaning is that he thought mistaken. For example, between the power of mental images to mislead one into thinking one means something and the view that meanings just are mental images.[17] Or again, is there not a connexion between suggestions III and IV (that one is led into constructing meaningless locutions by such things as seeing false grammatical analogies and treating meaning as an 'atmosphere') and Wittgenstein's rejection of what is sometimes called 'compositionalism', the idea that the meaning of a sentence is in some way *composed* of the meanings of its constituent words?[18]

In fact it would be surprising if he did not think that errors about how language works could bring about the supposed errors with which we are concerned. But there is a difficulty. What is the nature of these mistaken views about language? Are they supposed to be nonsensical or false? I doubt whether he thinks they are just false, but it seems circular to appeal to a nonsensicalist claim to justify nonsensicalism. Certainly someone who is not already a nonsensicalist will be unimpressed. I suppose that some nonsensicalists will say that errors about what meaning is and about

---

16. Or enough of it to avoid making all our talk of the mental seem suspect. I am aware, by the way, that many scientifically-minded philosophers are contemptuous of what they call 'folk-psychology', but would they be happy to see it as nonsense rather than a body of meaningful but woefully inadequate theory?

17. It would be too simplistic to say that philosophers wrongly credit themselves with meaning something because they are wrong about what meaning is, as though they were merely guilty of a rather sophisticated malapropism or like amateur naturalists using an inaccurate field-guide. I don't think any nonsensicalist would expect conversion to a correct view of meaning to make one immune to IOMs. What may be true - the matter will be discussed in the next chapter - is that a correct view of meaning will show how there is room for IOMs.

18. See H-J. Glock, *A Wittgenstein Dictionary*, Blackwell, 1996, pp.86-89.

when one means something oneself are intertwined and mutually reinforcing but it is far from clear how this is going to work, given that it is nonsense - which is incapable of entering into logical relations - rather than falsehood that is in question. If the illusion of seeing meaning in certain mistaken doctrines about language could be explained using some of the other ideas in this chapter, then perhaps adherence (if that is the word) to these mistaken, i.e. nonsensical, doctrines could be appealed to to explain without circularity how certain further cases of IOM are possible. But to me the whole question of the *efficacy* of nonsense is obscure. False beliefs (about language or anything else) can lead to other false beliefs. But what can being the victim of an IOM lead to? And how does it have the effects it has? My guess is that, even if nonsensicalism proves defensible, this will be one of the last problems the nonsensicalist will solve.[19]

VIII. Illusions Of Use.

Given the close connexion Wittgenstein finds between meaning and use and his view that language can fruitfully be seen as an assemblage of activities, it is natural to ask whether IOMs could be seen as *illusions of use*. Could it be that when someone is mistaken in thinking he means anything by a locution, he is mistaken in thinking he is *using* it, *doing* something with it? There are many passages in Wittgenstein which suggest this. Louis A. Sass[20] gives a long list of passages in which Wittgenstein produces 'striking metaphors by which he expressed a certain notion of futility' - measuring one's height by putting one's hand on one's head (*PI* 279); trying to make a car go faster by pushing from inside (*BB*, p.71); engines idling (*PI* 132); sitting at an empty loom going

---

19. I have not in this section mentioned Wittgenstein's emphasis on the fact that ordinary language is difficult to *survey*. It seems highly relevant to the difficulty of producing exceptionless generalisations about language and accurate accounts of the use of particular words, but I do not see how it could help explain IOMs (except in so far as it might lead one back to suggestion III).

20. *The Paradoxes of Delusion: Wittgenstein, Schreber and the Schizophrenic Mind,* Cornell U.P., 1994, pp.73-5. I do not think he really addresses the problem of what the futility of metaphysical utterances consists in, at least in a way that explains how they can seem non-futile. Nevertheless, I am indebted to him for making it clear to me just how pervasive are these images of futility in Wittgenstein. Sass, it is worth noting, takes a stance on one of the issues discussed in the last chapter: he sees philosophers and schizophrenics as in some cases subject to the same sorts of linguistic illusion.

through the motions of weaving (*PI* 414); a clock with a dial connected to the pointer (*BB*, p.71) ; words looked up in the imagination (*PI* 265); the right hand giving the left hand money (*PI* 268); a wheel that turns without affecting other parts of a mechanism (*PI* 271). To these one might add: thinking 'one is tracing the outline of a thing's nature ... [when] one is merely tracing the frame through which one looks at it' (*PI* 114); a knob that looks as though it turns on a machine but is merely decorative (*PI* 270); and the pseudo-machine discussed at *PG* 194 and *Z* 248.

Is there an account of IOMs to be got out of these passages? One way of taking them might lead one back to suggestion III. What the nonsense-talker fails to do is to make a legitimate move in a language-game: by confusing two language-games he fails to make a move in either of them.

Or one might see the passages as vivid illustrations of suggestion V: that one can think a sentence is still somehow being used when it is uttered outside the contexts in which it does have a use.

But perhaps there is another way of looking at them. Sass speaks of the 'futility' of philosophical utterances. The idea behind these passages seems to be that philosophical utterances do not achieve anything: they are in some way empty, pointless or idle. But it is not obvious how to formulate any such claim in a way that is not patently question-begging. The claim that philosophical problems arise when language 'goes on holiday' or 'idles'[21] has always seemed to me to rule out philosophical problems as not really problems, philosophical contexts as not really contexts, by thinly (very thinly) disguised linguistic legislation and I have never seen an account of the claim that does much to dispel this impression. But perhaps what is meant is that it will *turn out* that philosophical utterances are always somehow futile, where this futility is not

---

21. *PI* 38, 132. Some commentators prefer the translation 'idles' for both passages
22. At *PI* 520 Wittgenstein remarks that 'when we are tempted in philosophy to count some quite useless thing as a proposition, that is often because we have not considered its application sufficiently'. Commenting on this, Hacker (*Analytical Commentary on the Philosophical Investigations*, Vol. IV, Blackwell, 1996, p.313) writes:
> ... e.g. 'Cogito ergo sum', 'Every rod has a length', 'I know whether I am in pain or not' or 'I perceive I am conscious'. Unlike 'Is has good', these look like sensible sentences. It takes careful examination of how these sentence-like formations might be applied to see that they are in fact nonsensical.

Are they being rejected only because they have no *non-philosophical* application? If not, what counts as an application? .

simply a matter of their being philosophical.[22]

What notion of futility should one appeal to then? It is not clear that showing that an utterance is futile or an empty gesture is ever a sufficient reason for saying that the utterer means nothing by it. Think of the non-philosophical contexts in which we call what someone says pointless or an empty gesture, where one speaks of idle words, idle threats or boasts for example. We meet a familiar problem: in order to show that what someone says is futile, it looks as though we will have to ascribe a meaning to it. If we can make *nothing* of it, we will not be able to describe it with any confidence as futile. It may be that Wittgenstein's *Tractatus* view that tautologies and contradictions are 'senseless' survives in the later philosophy as a tendency to deny meaning to utterances he believes are in some way empty. (I am thinking particularly of the remark at *BB*, p.71, that 'the solipsist's "Only this is really seen" reminds us of a tautology', which comes shortly after the images of the clock with pointer fixed to dial and the man pushing a car from inside.) But in the *Tractatus* he was careful to distinguish the 'senseless' from the 'nonsensical' (*TLP* 4.461, 4.4611).

However, although I have been unable to turn this idea of an illusion of use into anything that clearly counts as a separate suggestion about how IOMs might be possible, I am reluctant to abandon it completely. Wittgenstein's images of futility in philosophy are so frequent that they demand to be taken seriously. There may be forms of the idea that someone suffering from an IOM thinks he is achieving something when he is not that I have not considered. One will have to find some enterprise in which the speaker has failed, on the basis of which it can be non-trivially concluded that he has failed to speak meaningfully ('non-trivially' since the enterprise must not just be: speaking meaningfully). And of course it must be possible for the speaker to be unaware of his failure.

IX. What Other Suggestions Might There Be?

Consider the following passage from the *Blue Book*, pp.65-66:

> When I say 'Only this is seen', I forget that a sentence may come ever so natural to us without having any use in our calculus of language. Think of the law of identity, 'a=a', and of how we sometimes try hard to get hold of its sense, to visualize it, by looking at an object and repeating to ourselves such a sentence as 'This tree is the same thing as this tree.'

The gestures and images by which I apparently give this sentence sense are very similar to those which I use in the case of 'Only *this* is really seen'. (To get clear about philosophical problems, it is useful to become conscious of the apparently unimportant details of the particular situation in which we are inclined to make a certain metaphysical assertion. Thus we may be tempted to say 'Only this is really seen' when we stare at unchanging surroundings, whereas we may not at all be tempted to say this when we look about us while walking.)

Here is a passage in which Wittgenstein is concerned with the phenomenology of philosophical error, indeed with IOMs. Is there anything that helps explain how they might be possible? Apart from the reference to images (a suggestion already considered) I am not at all sure that there is. This may surprise the reader, given that the passage seems so carefully focused on IOMs. So let us go through the possibilities.

Even if a sentence comes to us naturally, how does this explain our seeming to see meaning in it if it has none? (Does it not seem more likely that its seeming meaningful is part of the explanation for its coming to us naturally?) Even if we try to understand a supposedly nonsensical locution by means of various gestures and exercises of the attention, how could this explain our having the illusion of success? Is repeating a meaningless locution over and over again supposed to induce an IOM? - If so, how?[23] And even if we do utter meaningless locutions more often in some situations than in others, this surely does not explain our thinking we mean or understand something by them.

I mention this passage because it is a clear example of Wittgenstein's concern with the phenomenology of philosophical error and I can imagine that someone who recalls it and other such passages[24] might feel that they *must* help us with our problem if only we knew how to use them. I would be interested to know if anyone does think that we have here anything that could show how IOMs might be possible. I do not say this in an ironic or combative spirit. Wittgenstein is a subtle thinker who often expresses himself obliquely. New interpretations are constantly being put forward of

---

23. Curiously, in the case of single words Wittgenstein points out that some people find that they seem to *lose* their meaning if repeated many times. (*PI* Pt. II, p.214)

24. E.g. the discussion of the 'visual room' at *PI* 398.

familiar passages in his works and neglected passages are suddenly 'discovered' and emphasised. So perhaps there is some way in which a passage such as this could help us with our problem. But I do not at present see how and, apart from the idea to be discussed in the next chapter, I have now considered all the suggestions I know.

Why Do These Suggestions Occur In Such Rudimentary Forms?

I have not tried to reach a final evaluation of the above suggestions, although in many cases the mere attempt to formulate them precisely has uncovered serious difficulties. But what are we to say in general about the way the problem has been handled? If one were to tell someone new to the problem that perhaps eight or nine suggestions about how IOMs might be possible could be extracted from the literature, he would probably conclude that the issue was familiar and widely discussed. Yet there is an important sense in which he would be wrong. It is not always clear that the possibility of IOMs is what is at issue in the passages I have adduced; it is often unclear how to formulate the suggestions so that they clearly bear upon our problem and are not subject to obvious objections; and it is *never* clear how the suggestions are supposed to relate to each other. There is therefore a sense in which the problem has not been taken very seriously. In fact a case could perhaps be made for the view that the possibility of IOMs has simply been assumed, taken as a datum, and that the passages I have adduced do not deal with the problem at all: they just specify conditions which facilitate, make more likely, a phenomenon whose possibility is not being questioned.

(In fairness to the nonsensicalist I ought to remind the reader of what I said in the first chapter: that those who do not use nonsensicalist methods do not seem to have taken the issue any more seriously. Do they deny the possibility of IOMs? If so, on what grounds?)

It is hard to avoid the conclusion that nonsensicalists have never really had any doubts about the correctness of nonsensicalism; and I do not just mean any *serious* doubts. I mean that there is little evidence of even slight misgivings, the salutary awareness that, however confidently one might be forging ahead, it is just possible that one is on the wrong road. Why should this be? One possibility is that mentioned in Chapter Four, that some nonsensicalists have not clearly thought through what their accusations involve. But this cannot be the explanation in Wittgenstein's case nor in that of some of his followers.

I shall try to shed some light on the position of Wittgenstein. Most of the suggestions I have been considering come in fact from him, from his later work. Yet he does not carefully formulate them so as to make it clear that they are supposed to show the possibility of IOMs nor relate them to each other nor give a fully worked out account of how, why and when IOMs occur. Why should this be? Perhaps he thought that *anything* that goes on when someone genuinely means something by what he says could also go on in IOMs, and that there is therefore no point in carefully anatomising the possibilities. This is the basis of what I believe to be the most Wittgensteinian account of IOMs—though I do not find it stated explicitly in his writings—and I shall discuss it next.

Chapter Eight: The 'No Introspectible (Phenomenological, Experiential) Difference' Account.

Consider the following passages from Wittgenstein:

PI 329 When I think in language, there aren't 'meanings' going through my mind in addition to verbal expressions: the language is itself the vehicle of thought.

PI 693 ... nothing is more wrong-headed than calling meaning a mental activity!

PI Pt. II pp.216-17 Someone tells me: 'Wait for me by the bank.' Question: Did you, *as you were saying the word*, mean this bank? - This question is of the same kind as 'Did you intend to say such-and-such to him on your way to meet him?' It refers to a definite time (the time of walking, as the former question refers to the time of speaking) - but not to an *experience* during that time. Meaning is as little an experience as intending.

But what distinguishes them from experience? - They have no experience-content. For the contents (images, for instance) which accompany them are not the meaning or intending.

PI Pt. II p.217 If God had looked into our minds he would not have been able to see there whom we were speaking of.

PI Pt.II p.218 Meaning is not a process that accompanies a word. For no *process* could have the consequences of meaning.

Z 16 The mistake is to say that there is anything that meaning something consists in.

It will be evident that these passages have something in common. Wittgenstein denies that, when one means something by what one says, this meaning consists in any experience one is having at the time, or any mental activity, or any process. He is however perfectly prepared to admit that typically when one uses words meaningfully one does have experiences - images, for example, frequently cross one's mind.[1] But such mental occurrences accompanying an utterance are neither necessary nor sufficient for meaning something by an utterance.

---

1. For example, *PI* 663 - 'If I say, "I meant *him*", very likely a picture comes to my mind, perhaps of how I looked at him, etc.; but the picture is only like an illustration to a story. From it alone it would mostly be impossible to conclude anything at all; only when one knows the story does one know the significance of the picture.'

Now this suggests to me an account of how IOMs might be possible that is arguably more Wittgensteinian than anything considered so far. Moreover, it helps explain why the individual suggestions considered in the last chapter are presented in so undeveloped a form. On the other hand, I have to admit that I know of no explicit statement of the idea. What I have in mind might be put as follows:

> Meaning is not an *accompaniment* to one's words. What determines whether or not one means anything by what one says is not what happens to be going through one's mind at the time. This in itself seems to allow for the possibility of IOMs - it creates the conceptual space for them. If we ask what deceives someone into thinking he means something by an utterance when really he means nothing at all, we can no doubt mention the images, feelings, associations, gestures, etc., accompanying the utterance. Indeed *anything* that occurs when people speak meaningfully - any accompaniments to the words, any experiences, processes, activities - could also occur when they falsely believe they are speaking meaningfully, since such accompaniments are neither necessary nor sufficient conditions for meaningful speech. But instead of focusing on particular mental phenomena and elaborating emotion accounts, mental image accounts, and so on, it is better just to stress that, since meaning is not a matter of 'introspectible entities', a person is not necessarily the final authority on whether he means anything.

I have placed 'introspectible entities' in scare-quotes because the very phrase might raise Wittgensteinian hackles, suggesting as it does some kind of interior perception. The point is that, however one should best describe such things as pains about which the sufferer *is* the final authority, one should not assimilate meaning something by what one says to them. It is even possible that someone proposing this account would regard excessive concentration on the kind of thing discussed in the last chapter as analogous to the mistake mentioned at *PI* 314:

> It shews a fundamental misunderstanding, if I am inclined to study the headache I have now in order to get clear about the philosophical problem of sensation.

The idea would be that 'introspection' is never the way to approach philosophical problems, and that, even when explaining philosophical error, one should be wary of it. But, without going this far, we seem to have found an alternative approach to the problem of IOMs, one which de-emphasises the experiential or any detailed examination of the experiential and merely says that what determines whether someone means anything by his words is not what is going on in his mind at the time.[2]

Wittgenstein in fact shows considerable hostility to questions of the form 'What goes on when ...?', asked about the psychological. We can see from what has already been said why he would have been suspicious of the question 'What goes on when someone *means* something by what he says?' But I do not know that he ever takes this line with the question 'What goes on when someone *wrongly thinks there is something he means* by what he says?' To do so might easily convey an impression of evasiveness. It is better, I think, to say that anything that *goes on* when there is something one means could also *go on* when one only thinks there is. Z 88 is one of his dismissive remarks about 'what goes on' - 'It is very noteworthy that *what goes on* in thinking practically never interests us. It is noteworthy but not queer.'[3] I am half-inclined to retort, 'Speak for yourself.' Nevertheless, it might be that it is easy to give the wrong significance to what goes on. The present account could be seen as a Wittgenstein-inspired attempt to give the right significance to what goes on both when one speaks meaningfully and when one merely thinks that one does.

This is not an *interpretation* of Wittgenstein in the usual sense but an attempt to apply some of the things he says about meaning to the question of how IOMs might be possible in a way that would explain why his specific comments on the sorts of thing that mislead

---

2. Not all the suggestions considered in the last chapter, it might be said, seem to have much to do with the introspectible or experiential. But this may be because of their undeveloped nature. If one conceives one's task as explaining how in a particular case someone is deceived into thinking he means something when he does not mean anything, one will surely at some point have to consider the *phenomenology* of the deception. The only way to avoid incurring this obligation that I can see is by offering an account like the present one: phenomenologically, genuine meaning and IOMs can be as similar as you please, so the details of any particular case will matter less. As I explain at the end of the chapter, giving this account puts one under a different obligation.

3. The explicit concern here is with *thinking* but I feel confident that he would have said the same about what goes on when one speaks meaningfully. See David G. Stern, *Wittgenstein on Mind and Language*, O.U.P., 1995, pp.105-7, for a discussion of Wittgenstein's hostility to 'What goes on when ...?' questions.

philosophers are not elaborated into a *theory* of IOMs.[4] Since he never, to my knowledge, asks the question, 'Is it actually possible to be mistaken in thinking one means anything by what one says?', it is perhaps too much to expect a direct answer to it. I could not with any confidence claim to be expounding *his view*. The following passage (*BB*, p.65) is suggestive however:

> Now if for an expression to convey a meaning means to be accompanied by or to produce certain experiences, our expression may have all sorts of meanings, and I don't wish to say anything about them. But we are, as a matter of fact, misled into thinking that our expression has a meaning in the sense in which a non-metaphysical expression has ... The meaning of a phrase for us is characterised by the use we make of it. The meaning is not a mental accompaniment to the expression.

If we assume that he means that the meaning of a phrase actually is 'characterized by the use we make of it' - not just 'for us' - then it is possible to see him as saying that anyone who (wrongly) thinks that the meaning is a mental accompaniment to the expression will also be liable to treat people, himself included, as the final authority on whether they mean anything. On this view it is not so much that an expression might have 'all sorts of meanings' as that all sorts of things can help delude one into thinking there is something one means when there is not. We have in fact already noted that there seem to be parallels between what Wittgenstein considers to be mistaken views about what meaning is and the sorts of thing he believes delude people into mistakenly thinking there is something they mean.

One merit of this account is that it goes some way towards explaining what one might call the *smoothness* of the deception (supposing it to be a deception) effected by 'philosophical nonsense'. Wittgenstein often appears to be discussing cases where philosophers either seem uncertain whether they mean anything or feel the need to affirm that

---

4. I do not think it sufficient to attribute this to his stylistic idiosyncrasies nor to his dislike of systematic philosophy. The idea of IOMs is so important to his general approach that he surely needs to state clearly why he thinks they are possible and the fact that he does not demands an explanation. Nor do I think that his view that philosophers should not produce *theories* is relevant here. We require a theory only in the sense that an account of how self-deception is possible might be called 'a theory of self-deception'.

they do. (Immediately after the passage just quoted he speaks of philosophers who say, 'I think I mean something by it' or 'I'm sure I mean something by it.') He often imagines philosophers trying to 'give sense' to expressions. But until philosophers started trading nonsensicalist accusations it would surely rarely if ever have crossed a philosopher's mind that he meant nothing by what he said. Recall the following remark:

> Think of the law of identity, 'a=a', and of how we sometimes try hard to get hold of its sense, to visualize it, by looking at an object and repeating to ourselves such a sentence as 'This tree is the same thing as this tree.' (*BB*, p.66)

I suspect that a philosopher who does this already has one foot in the nonsensicalist camp or has been accused of talking nonsense by others. He seems to have acquired nonsensicalist scruples. But the metaphysician who thinks 'A thing is identical with itself' is a necessary truth, which is far more certain than more specific identity statements, such as 'Hesperus is Phosphorus', will not normally feel the need for any such mental exercises. Now the account of IOMs that we are considering has an explanation for this: someone proposing it might say, 'Just so; there is a sense in which IOMs are *exactly like* genuine cases of meaningful utterance.'

I suggest therefore that if the later Wittgenstein had been asked to say why he believed it possible to be mistaken in thinking one means anything, he might have given an account like the one I have sketched. I am unable to be more definite in attributing this view to him. To claim that it was anything more than 'at the back of his mind' would require that one make some attempt to answer the following questions at least:

a) Why does he never state it explicitly?

b) How is one to reconcile the numerous cases where he does seem to be trying to give phenomenological accounts of how the deception in IOMs is effected with a general account of IOMs which seems to suggest that there is little point in doing so?

c) What would he have said in his *Tractatus* days, long before he had formulated the views on what meaning is not on which this account depends?[5] In what follows I shall simply treat the account as one that is clearly inspired by certain things in the later Wittgenstein and try to consider it on its own merits.

Is This Way Of Explaining IOMs Circular?

The passages quoted at the beginning of the chapter were mainly about what meaning is *not*. So what is the status of such views as that meanings are mental images or processes accompanying words? If they are supposed to be merely false there is no problem, but I doubt whether Wittgenstein thinks this. If they are supposed to be nonsensical, then the explanation seems to be appealing to nonsensicalism in order to justify nonsensicalism. How damaging is this apparent circularity?

Freudians, Marxists, Christians, Existentialists are all at times inclined to use their own doctrines to explain why others refuse to accept those same doctrines. A person's unconscious resistances, class position, sinful nature, or bad faith may be cited as the real reason why someone cannot - or will not - see their truth. Such claims are rightly viewed with suspicion. But one needs to be clear just what is suspect about them. So far as I can see, they do not involve any inconsistency. For example, assuming it is at least possible for someone's unconscious desires or fears to prevent his giving a fair hearing to a certain theory, could not this happen when the theory in question is Freudianism? The real difficulty is that it is hard to see why such an accusation should cut much ice with the accused: it seems *strategically* inept. He has, presumably, conscious reasons for rejecting Freudianism and he can surely expect a Freudian to engage with those reasons. If he later comes to think of them as weak, *then* he can look for unconscious determinants of his earlier intransigence.

---

5. Cora Diamond ('Ethics, Imagination and the *Tractatus*' in *The New Wittgenstein*, ed. Alice Crary and Rupert Read, Routledge, 2000, p.159) sees the early Wittgenstein as taking over Frege's uncompromising anti-psychologism. This might, if consistently worked out, lead to something like the later Wittgenstein's denial that meaning is an experience or process accompanying one's words. Certainly Diamond thinks so. But without wishing to deny that the *Tractatus* is resolutely anti-psychologistic, I have to say that I do not think she provides much textual evidence that the early Wittgenstein did pursue this line of thought. There is surely a limit to what one can legitimately *read into* a philosopher's writings and I think I have reached that limit by describing the present account of how IOMs might be possible as late-Wittgensteinian in inspiration.

THE 'NO INTROSPECTIBLE DIFFERENCE' ACCOUNT

How does this apply to our problem? There does not seem to be any obvious inconsistency in the claim that a philosopher might be taken in by a nonsensical claim about meaning (i.e. he is mistaken in thinking he means anything by it) and this leads him to reject or doubt the possibility of IOMs (*How* it does so will need to be spelt out and this will not be straightforward: we are dealing with nonsense, not falsehood, so it cannot be a matter of following genuine logical implications.) It seems at least conceivable that his scepticism about IOMs might somehow be the result of his being the victim of IOMs.

But there are two difficulties. First, *I* am sceptical about IOMs but I do not think that this suggestion applies to me. Certainly I do not profess any of the views about meaning which we saw Wittgenstein rejecting at the beginning of the chapter (though I believe that if nonsensicalism proves untenable, then all his arguments and conclusions concerning meaning will need re-examining to see whether they can be stated without recourse to nonsensicalism). So the claim would have to be that the mere fact that I do not *reject as nonsensical* these views about meaning leads me into scepticism about IOMs. That I find hard to swallow. I want to reply that it just is unclear whether IOMs are possible and any reluctance to confront the problem is evasive.

This leads straight into the second difficulty, a strategic difficulty. Why should anyone who is not already a nonsensicalist be impressed? We have by now uncovered a fair number of difficulties for nonsensicalism and it seems reasonable to demand that they be overcome in ways that do not themselves assume its truth. If this is achieved, it will be appropriate then and not before to ask whether any resistance to the idea of IOMs was due to the influence of IOMs.

I can however imagine the following reply:

> The difficulty that the present account of IOMs is supposed to deal with is that it is not clear that they are possible; let us leave any other difficulties for the moment. The present account does deal with this. True, it may have to appeal to IOMs at some point but given that the objection it is designed to meet is of the form 'I don't see how ...' rather than 'IOMs are impossible because ...', this may not be too serious. If you are willing to accept that nonsensicalism *might* be correct, you can consider this account and see whether it is consistent and works. You admit it is consistent. Does it not also work?

DO PHILOSOPHERS TALK NONSENSE?

Perhaps it does. From the nonsensicalist standpoint it looks as though we have an explanation of how IOMs are possible. They are possible because what determines whether someone means anything by an utterance is not what is going through his mind at the time. Views which apparently deny this may themselves be the result of IOMs (Even the difficulty one has in seeing how IOMs are possible might be the result of one's susceptibility to IOMs but, as I said, I find that implausible.) The fact that I can adopt the nonsensicalist perspective and see the merits of this account makes me reluctant to dismiss it on grounds of circularity alone, though I am also reluctant to take up residence in the nonsensicalist camp.

Clearly the difficulty we are considering is likely to arise however the nonsensicalist tries to demonstrate the possibility of IOMs. Unless he can base his account entirely on claims he thinks could *meaningfully though falsely* be denied, there will be a suspicion that he is using nonsensicalism to justify itself.[6] The same problem would have arisen with some of the suggestions in the last chapter (see in particular section VII). However, as I hope the above discussion has shown, it is far from obvious how serious the difficulty is. Perhaps the real problem is whether there is any way in which someone not already a nonsensicalist can be rationally induced to become one. Would his conversion be anything more than a leap of faith? This question is closely related to a more straightforward difficulty for the present account of IOMs than that of apparent circularity.

If What Decides Whether Someone Means Anything Is Not What Is Going On In His Mind, What Does Decide?

Suppose one were to accept this account of how IOMs are possible. If one wanted to judge the success of a particular nonsensicalist accusation, everything would then depend upon what grounds the accuser could produce for saying that the accused did not mean anything.

Consider an analogy. A counterfeit banknote can be indistinguishable

---

6. There is a general problem about interpreting Wittgenstein lurking in the background here. He (Preface to *PI*, p.vii) speaks of himself as travelling 'over a wide field of thought criss-cross in every direction'. This makes it difficult to know what he is entitled to assume at any particular point in his discussion. Perhaps he would protest that he does not make assumptions any more than he propounds theses and theories. But are there not passages in which he just is assuming the possibility of IOMs? If so, the question will arise whether he already said anything to entitle him to this assumption.

from a genuine one and yet still be counterfeit. Why? Because what determines whether it is genuine is whether it was issued by the accredited authorities and this cannot necessarily be decided by inspection. Now we are considering the view that a case in which someone mistakenly thinks there is something he means by what he says and one in which he correctly thinks there is might be indistinguishable in respect of what is going on in his mind at the time - indistinguishable by *intro*spection, one might say provided one heeds the warning given earlier. So something else must make the difference. What? (Obviously it is not likely to be anything similar to what makes the difference between counterfeit and genuine banknotes.) If *illusions* of meaning can perfectly mimic genuine cases, this raises the question: why speak of illusions at all?

It is worth noting that the same question would have arisen with many of the suggestions in the last chapter, if we had evaluated them individually. The mental image account, for example, would have prompted the following query:

> But mental images often accompany genuine cases of meaningful speech, as Wittgenstein for one freely admits. How are we to know whether someone is being misled by a mental image into wrongly thinking there is something he means by an utterance or whether we have a harmless case of an image that accompanies the meaningful use of a locution?

Chapter Nine: What Grounds Could One Have For Overruling Someone's Sincere Claim To Speak Meaningfully?

The last chapter left us with the question whether one could ever have grounds for saying to someone, 'Although you genuinely believe you mean something by "*p*" and, although anything that goes on in the mind of someone who really does mean something by what he says may well be going on in yours, you do not in fact mean anything by "*p*". There is another reason why we need to ask this. In Chapter Three I drew attention to the Problem of Specifying the Nonsense but have not returned to it since. I now want to consider briefly a way in which the nonsensicalist might try to deal with that problem, one which has some plausibility but which immediately raises the question, 'Why speak of nonsense?'

When discussing the notions of 'misleading grammatical analogies', 'language-game conflations' and 'category-mistakes' in Chapter Seven, I expressed a doubt whether they could - at least on their own - explain how someone could mistakenly think that he himself meant something by a locution as distinct from merely thinking that it had meaning. It might seem however that they could more successfully be used to deal with the Problem of Specifying the Nonsense.

Suppose two philosophers are arguing about whether X is or is not identical with Y. (We could equally well take some other kind of dispute - whether X is a member of class C, say, or whether it has property P - but let us stay with identity statements.) A nonsensicalist intervenes, claiming that it is *nonsense* to identify X with Y. To avoid the 'sense that is senseless' difficulty he must show that he is not giving a meaning to 'X is identical with Y' and then claiming that, so interpreted, it is meaningless. What he does is this: he compares 'X is identical with Y' with identity statements involving X or Y which he thinks do make sense and claims (a) that none are genuinely analogous to 'X is identical with Y' but (b) the disputants have failed to realise this and (c) this has misled (or at least helped to mislead) them into thinking they mean something by it. He might say:

> Obviously you could give 'X is identical with Y' a sense by giving one or more of the words in it a new sense but you have not done so and would no doubt think it irrelevant to do so. You think you are using words in their normal senses at

least in part because you are misled by false analogies. There is no meaningful way of using 'X is identical with Y' on analogy with 'X is identical with Z' (something that does make sense).

This may work.[1] Provided the nonsensicalist avoids saying or implying that the accused *is* identifying X with Y, provided he confines himself to saying that the accused *thinks* he is making a move in a language-game analogous to that made when one identifies X with Z, or that he *tries*[2] to make such a move, or that he *seems* to see a sense in 'X is identical with Y' analogous to that of 'X is identical with Z', he may have successfully specified the nonsense he is rejecting. He has not just drawn attention to the wholly contingent fact (supposing it to be a fact) that 'X is identical with Y' is not part of the language but has managed to specify what sort of mistake the person who seems to see a sense in it is making. Looking at the matter in another way, we have here an account of what kind of understanding the nonsensicalist claims to have of those he accuses, of his implicit claim to understand them better than they understand themselves.

But what justifies these somewhat tortuous manoeuvres? Why deny in the first place that the accused means anything by the locution? If it has already been decided that he does not, this

---

1. I have used late-Wittgensteinian terminology in formulating this suggestion. The reader might like to consider for himself how it might be brought into relation with the early Wittgenstein's reflexions on Russell's Theory of Types (with which I introduced the Problem of Specifying the Nonsense in Chapter Three) and with Ryle's notion of a category-mistake. I ought perhaps to mention, since I criticised Malcolm's *Dreaming* for its lack of sophistication in the handling of the concept of nonsense, that even there there is a hint of the idea I am considering. On p.6 he says that ' "I am asleep" does not have a use that is homogeneous with the normal use of "He is asleep" '.

2. Edward Witherspoon in 'Conceptions of nonsense in Carnap and Wittgenstein' in *The New Wittgenstein*, ed. Alice Crary and Rupert Read, Routledge, 2000, pp.342-4, argues that an apparently similar appeal to trying does not work because it only makes sense to speak of trying to X if it makes sense to speak of X-ing. I am not sure what to say about this. I am imagining a nonsensicalist who appeals to the the notion of someone's trying to treat the grammar of one word as analogous to that of another and, because they are not analogous, failing to speak meaningfully. It is simply not obvious to me whether Witherspoon's strictures apply here. If they do, I doubt whether the 'Carnapian' conception of nonsense he is criticising will be the only casualty. It might turn out that one cannot specify the nonsense in question even via the mistakes made by the nonsense-talker; and then will not all nonsensicalists be stymied? However here I am interested in asking what our response should be if something like the present suggestion can in fact be made to work.

may be the way to get round the 'sense that is senseless' difficulty, but *how* has it been decided? Why say the analogies influencing the accused are misleading ones? Why talk of merely 'thinking', 'trying' or 'seeming' instead of allowing that the accused has (truly or falsely, but intelligibly) identified X with Y? Again, we must ask what grounds there could be for overruling a sincere claim to speak meaningfully.

But first I would like to discuss a passage which helps to bring out the difficulty of the problem. Edward Witherspoon's defence of an austere conception of nonsense has already been mentioned several times and from it I have learnt a great deal. But at the end of his paper he briefly discusses how, if one does not assume that the words in a suspect utterance have their normal meanings which somehow combine to produce nonsense, one can nevertheless come to the conclusion that the utterance is nonetheless nonsense. He writes:

> ... when Wittgenstein is confronted with an utterance that has no discernible place in a language-game, he does not assume he can parse the utterance; rather, he invites the speaker to explain how she is using her words, to connect them with other elements of the language-game in a way that displays their meaningfulness. Only if the speaker is unable to do this in a coherent way does Wittgenstein conclude that her utterance is nonsense: ideally, the speaker will reach the same conclusion in the same way and will retract or modify her words accordingly.[3]

I realise that this is intended as no more than an *indication* of how a Wittgensteinian nonsensicalist ought to proceed but it seems to me that nothing along these lines is likely to work. There is surely a variety of reasons which Witherspoon does not consider why the speaker might be unable to explain how she is using her words in a way that satisfies her interrogator. It could be the difficulty one regularly encounters in explaining philosophically interesting concepts. Few today would accept that Socrates was right to assume that if one cannot *define* the words one is using, one does not know their meaning. Yet nonsensicalists are apt to assume that, in cases where there is a real possibility that someone is not using words in the normal way, failure to give a precise explanation of how they are

---

3. Ibid., p.345. The assertion that this represents Wittgenstein's actual practice seems to me to be quite unjustified. I shall treat it as a proposal about how the Wittgensteinian nonsensicalist *ought to* proceed.

being used shows that nothing is meant by them. Or the explanation could be (somewhat uninterestingly, I admit) the obtuseness of the interrogator. Or, as will emerge in the course of this chapter, it could be that the interrogator is imposing conditions on what counts as an acceptable explanation which she does not accept and which perhaps have no general validity. How are these possibilities to be eliminated?

If it is unclear how this kind of interrogation could reveal to the interrogator that the explanation of her inability to explain her words is that she means nothing, it is equally unclear how it could reveal this to the speaker herself. Witherspoon notes that she might 'modify' her words, which suggests that she might acknowledge that she has expressed herself badly, but this is not what a nonsensicalist wants to establish.

Let us now proceed to a more systematic consideration of how a nonsensicalist might try to show that someone is not speaking meaningfully. I shall consider the following possibilities:

a) that there might be criteria of meaningfulness;

b) that one might be able to make 'disguised' nonsense 'patent';

c) that there might be some direct way of showing someone that he has given no meaning to a certain locution;

d) that, in order to show someone that he means nothing, one might require the co-operation of that person.

a) Criteria Of Meaningfulness.

It should be clear from our discussion of Malcolm's verificationism and the Problem of Diagnosis what the difficulty with criteria of meaningfulness is going to be: namely, how to apply a criterion without granting meaning to whatever utterance is under consideration. It is not obvious how there can be such a thing as a criterion of meaningfulness. Nevertheless, when the point is made at so general and abstract a level it is difficult to be sure that there is not some possibility that one has missed. Added to this is the problem that there is no obvious way of listing all the criteria of meaningfulness that might conceivably be proposed. I shall confine myself to considering whether one could modify verificationism so as to avoid according meaning to the very utterance one wants to claim has none. This should emphasise the seriousness of the difficulties that are likely to confront *any* proposed criterion of meaningfulness.

DO PHILOSOPHERS TALK NONSENSE?

I mentioned in Chapter Two that Malcolm sometimes applies his verificationist demand to the question whether someone understands a claim or uses it correctly:

> It is logically impossible that there should be a criterion for saying that someone understands how to use the sentence 'I am asleep' to describe his present state. This is equivalent to saying that the idea of such a use is not intelligible.[4]

One could not teach someone the use of 'I am asleep' if this requires that one draw his attention to the fact that he is asleep and tell him that this is a suitable occasion for using the locution or that he should exhibit his understanding of it by using it only when he is asleep. Many questions have been asked about this argument. Is it legitimate to make such a demand of individual sentences rather then to assume that if a person has learnt the use of 'I' and 'asleep' in other sentences, he can extrapolate to 'I am asleep'? Why can one only exhibit one's understanding of sentences by using them when they are true? Is Malcolm perhaps assimilating sleep to unconsciousness, coma, being 'out cold'? But here I want to ask whether Malcolm, by focusing on teaching and checking correctness of usage, has succeeded in distancing himself from the suspect locution so as to avoid the 'sense that is senseless' difficulty. I cannot see that he has. Does he not still have to understand 'I am asleep' to see the difficulty (great or slight) of knowing 'that someone understands how to use the sentence "I am asleep" to describe his present state'? It is as though he allows himself to understand *what it would have to mean* and then claims that neither it nor anything else could mean that.

What might seem a more promising modification of verificationism could be called 'laid-back verificationism'. Suppose that a philosopher, instead of undertaking to prove that a claim is unverifiable, simply says that he does not understand it unless you tell him how to verify it.[5] This does sometimes happen, especially with Wittgensteinians who deploy the term 'criterion': they sit back

---

4. *Dreaming*, pp.16-17.
5. Some of the early versions of the verification principle do seem to legitimise this, e.g. 'The sense of a proposition is the method of its verification. A method of verification is not the means of establishing the truth of a proposition, it is the very sense of a proposition ... You cannot look for a method of verification.' (*Ludwig Wittgenstein and the Vienna Circle - Conversations recorded by Friedrich Waismann*, ed. Brian McGuinness, tr. Joachim Schulte and McGuinness, Blackwell, 1979 p.227)

and demand the criterion for such-and-such.[6] This is not so much the employment of a criterion of meaningfulness as a demand for the criterion for the truth of a particular utterance with the implication that it is meaningless if no criterion is forthcoming. And here, I think, lies the difficulty. If the laid-back verificationist confines himself to saying that *he does not understand* the utterance for which no method of verification has been given (no criterion provided), then I do not see that anything can touch him (even if one suspects him of disingenuousness). But if he goes beyond the autobiographical and claims that *the speaker is mistaken in thinking he means anything*, either he will have to show that there *could* be no method of verification, in which case we are back with ordinary verificationism, or he will have to claim that the mere fact that the speaker cannot say how it could be verified shows that he means nothing by it. This latter claim seems horrendous: surely we allow that someone can conceive of a possibility *and then* ask whether and how one could tell whether that possibility obtained. Any suggestion that one somehow only comes to mean anything by one's supposed description of the possibility once one has thought of a means of verification (as implied by the passage quoted in footnote 5) can only be countenanced on the charitable assumption that 'mean' is being used in a new technical sense. The 'sense that is senseless' difficulty has not been left far behind, for one would surely have to mean something (in the ordinary sense of 'mean') by the description in order to set about trying to think of a method of verification.

Another possible way of refurbishing verificationism is suggested by Flew's attack on theism.[7] He holds that theists are wont to drain their own thesis of meaning by refusing to allow anything to count against it. Thus, any empirical evidence against the view that the world was created by an omnipotent, omniscient, benevolent being, is always somehow explained away: it becomes unclear what content theism has. This might seem an improvement on straightforward verificationism in that theism presumably *began* as a claim with

---

6. We saw in Chapter Two how Hacker took this line about Luria's mnemonist.

7. 'Theology and Falsification' in *New Essays in Philosophical Theology*, ed. A.G. N. Flew and Alasdair MacIntyre. Macmillan, 1955. Flew's argument stimulated a large literature. Here I am concerned only with whether it can be seen as a successful attempt to circumvent the general difficulty facing criteria of meaningfulness that we are considering. I do not think the shift from verification to falsification introduces anything importantly new.. Blackwell, 1979 p.227)

empirical implications but in response to criticism was gradually watered down by the theists themselves. Flew could say that he never tries to show that theism is unverifiable (or unfalsifiable); rather he treats it as having empirical implications but finds that the theists themselves discount these implications, leaving him wondering whether they are really committing themselves to anything.

Whilst I agree that at least some theists do lay themselves open to this charge, I doubt whether it is best seen as a nonsensicalist one. For do not theists have to mean something by their claims if they are to react in the way that Flew alleges? How otherwise would they recognise potential conflicts with features of the world and how would they know what to do to explain them away? If it is replied that this only applies to the early stages of the slide into vacuity, I would ask how it is known when total vacuity is reached and whether the theist does not have to retain some grasp of his watered-down thesis in order to check that it is indeed compatible with whatever evidence is brought against it. Perhaps he is better seen as perpetually shifting his ground. Or perhaps as illustrating Quine's dictum that '[a]ny statement can be held true come what may, if we make drastic enough adjustments elsewhere in the system.'[8]

Although I have confined myself to verificationism in the above, it should be clear how problematic any criterion of meaningfulness is going to be. One possibility that has emerged and which will need to be further investigated is that the nonsensicalist is using the word 'mean' non-standardly. Even if it is possible to avoid the 'sense that is senseless' difficulty by saying, 'A person does not mean anything by an utterance unless he can explain it in accordance with the following conditions ...' (there is the possibility here that non-verificationist conditions will be suggested), we must ask whether this notion of what it is to speak meaningfully is the everyday one.

b) 'Disguised' To 'Patent' Nonsense.

I now wish to return to a question raised in Chapter Three. At *PI* 464 Wittgenstein writes, 'My aim is: to teach you to pass from a piece of disguised nonsense to something that is patent nonsense.' Elsewhere[9] he remarks that 'operations' are needed to effect this.

---

8. *From A Logical Point Of View*, Harvard U.P., 1980, p.43.
9. *Wittgenstein's Lectures, Cambridge 1932-1935, from the notes of A. Ambrose and M. Macdonald*, ed. Ambrose, Blackwell, 1979, p.64.

What operations? One cannot transform the disguised nonsense into something that means the same, since it is not supposed to mean anything. But neither can one ask what follows from it. One can show that something is false by showing that it entails a falsehood but not that it is nonsense by showing that it entails nonsense. So what is left?

The only thing I can think of is that one might try to exhibit the nonsensicality of something by *comparing* it to something that is patent nonsense. But this procedure is likely to have little cogency. The two locutions will be different, so how can one be sure that one difference is not that one makes sense and the other does not? One might also find that what is thought patent nonsense by one philosopher is not thought so by another. Finally, even if one could show by this method of comparison that a locution is excluded (though not obviously) from the language, this would not show that its utterer meant nothing by it. So the operations needed to show that someone is talking nonsense must involve the utterer as well as the utterance. It is unclear what they could be.

Consider an example. Hacker and Baker cite a manuscript of Wittgenstein's in which 'A thing is identical with itself' and 'A thing is very similar to itself' are given as examples of disguised and patent nonsense respectively.[10] Now to claim that the former entails the latter would be to make the mistake of treating nonsense as having entailments. So presumably it is the resemblance of the former to the latter that is held to condemn it. But could it not be that one makes sense whilst the other does not? After all, nonsensicalists themselves tell us to beware of grammatical resemblances! And is it so obvious that 'A thing is very similar to itself' is nonsense? It is unfamiliar to philosophers for, unlike 'A thing is identical with itself', it has not been allotted a leading metaphysical role. But if someone were to claim that it is perfectly meaningful, though embodying a less important metaphysical truth than its illustrious relative, it is not clear how one could dislodge him. Finally, even if one could show that 'A thing is identical with itself' is, contrary to appearances, not a grammatical English sentence, this would not show that philosophers who utter it mean nothing by it.[11]

The underlying difficulty is surely that, as with criteria of

---

10. *Analytical Commentary on the Philosophical Investigations*, Vol. II, Blackwell, 1985. p.208.

11. In his exegesis of *PI* 464 (*Analytical Commentary*, Vol. IV, Blackwell, 1996 pp.142-3) Hacker provides an interesting discussion of possible examples of making disguised nonsense patent but in each case it seems to me that he either treats nonsense as having entailments or uses the method of comparison without noticing its inconclusiveness

meaningfulness, one is trying to get to work on an utterance in order to show something significant about it but without according meaning to it.

c) A Direct Pointing To Meaninglessness.
At *TLP* 6.53 Wittgenstein writes:
> The correct method in philosophy would really be the following: to say nothing except what can be said, i.e. propositions of natural science - i.e. something that has nothing to do with philosophy - and then, whenever someone else wanted to say something metaphysical, to demonstrate to him that he had failed to give a meaning to certain signs in his propositions.

Could there be some direct way of drawing someone's attention to the fact that he had given no meaning to certain signs, some Zen-like nudge that would draw his attention to *an absence*? The last two ideas ran into difficulties and it might well seem that this is the only possibility left. Perhaps we have still not fully woken up to the fact that it is *nonsense* we are supposed to be talking about and we persistently try to use methods which are adapted to bringing out defects in utterances which at least mean *something*[12]; what we must do is to recognise that nonsensicality is meaninglessness and that the only way of exposing it is to show that there is just no meaning there.

But this is easier said than done. How is one to draw someone's attention to this absence of meaning? Or how is he to become aware of it unaided? He is supposed somehow to become aware that he means nothing by a locution.[13] Does this ever happen? The most plausible case is perhaps that of the drug-taker in Chapter Six, who

---

12. Perhaps one could say that one cannot have *reasons* for calling something nonsense other than, trivially, that it does not mean anything nor for saying that someone is talking nonsense other than, trivially, that he does not mean anything. Once one starts saying 'It is nonsense because it is ...' one is already in difficulties.

13. The way I am using *TLP* 6.53 might be questioned. Those philosophers who see Wittgenstein as distinguishing 'illuminating' from 'mere' nonsense might wonder whether someone trying to say what can only be shown, i.e. producing 'illuminating' nonsense, does mean something by what he says even though he is somehow unable to give meaning to the signs by which he tries to express his meaning. But even if this is right, 6.53 is surely meant to apply to 'mere' nonsense as well as 'illuminating' nonsense. There is no suggestion that *all* metaphysics is an attempt, worthy of respect, to say what can only be shown.

seemed to see meaning in some bizarre locution but ceased to do so once the drug's effects had worn off. (Even there it was not clear how to eliminate the possibility that he had simply forgotten what he meant.) But what is striking is the contrast with the philosophical case, where typically locutions that a philosopher has decided are philosophical nonsense still *seem* meaningful to him. It looks as though Wittgenstein's would-be metaphysician will have to realise that he has given no meaning to certain signs in his 'propositions' in spite of their still seeming meaningful to him. But even if this happens, why should one trust the supposed direct awareness of the absence of meaning rather than the persisting appearance of meaning? A nonsensicalist might claim that if someone did fully realise that he had given no meaning to the signs in question, then *all appearance* of meaningfulness would vanish. I would question whether this has ever happened in philosophy and whether there is any warrant for thinking it ever will.

Our discussion in Chapter Four of the nature of nonsense suggests that we may not have been formulating this idea of a direct pointing to meaninglessness entirely accurately. Cora Diamond, it will be remembered, argued that for Wittgenstein there is no 'positive nonsense'. One cannot combine meaningful locutions to get nonsense - that is, the locutions do not retain their meaning in the nonsense. Now suppose we have what appears to be a subject-predicate proposition 'S is P' and it somehow transpires that we have given no meaning to 'S'. We could equally well say that we have given no meaning to '... is P' in this sort of linguistic environment; the fact that there may be legitimate uses of '... is P' is irrelevant. So it is not that 'S is P' splits up into a meaningless and a meaningful part; it is rather that the whole thing is nonsense. So perhaps neither we nor Wittgenstein should have spoken of someone's failure to give a meaning to certain signs in his 'propositions', as though some terms in them did have meaning. Rather, all his metaphysical 'propositions' would be meaningless throughout.[14]

I do not think that this reformulation makes it any easier to see how someone could simply realise that he is speaking without meaning anything. It may even make it harder. The person will not be able to focus on some particular sign and say, 'Aha, I have let that slip past

---

without giving it a meaning.' He will somehow have to come to see his putative metaphysical propositions as meaningless from beginning to end.

I would like to end this section by discussing a worry that the reader may have had: is there a danger that our discussion is becoming *too epistemological*? It is natural to formulate the Problem of Diagnosis as: How could one ever *know* that someone was suffering from an IOM? This way of looking at the matter does bring out the difficulty of being a consistent nonsensicalist but it also raises the question of what the status of IOMs is, if they are undiagnosable, undetectable. We could hardly respond by becoming verificationists and dismissing the hypothesis of IOMs as nonsensical. Ought we therefore to allow the theoretical possibility of IOMs but maintain that one could never do more than *suspect* that someone was the victim of one? That seems consistent but unsatisfactory. I will not attempt a final evaluation of nonsensicalism until the last chapter but one point can conveniently be made here.

I opened the chapter by imagining a nonsensicalist who approached the Problem of Specifying the Nonsense by speaking of someone as, for example, 'seeming to see a sense' in 'X is identical with Y' analogous to the (genuine) sense of 'X is identical with Z' or as thinking he is making a move in a language-game analogous to that made when one identifies X with Z. What our discussion strongly suggests is that such manoeuvres are *pointless* unless reasons are given for refusing to allow that the accused is speaking meaningfully when he says, 'X is identical with Y' and that suitable reasons are going to be hard to find. Why are these elaborate formulations more accurate than a simple ascription of meaning to the utterance? Why is he not simply identifying X with Y (rightly or wrongly)? In the next chapter I suggest that typically what is really going on is that the nonsensicalist employs the ordinary concept of meaning when investigating an utterance, probing it, and then concludes that it lacks meaning in some narrower sense of his own.

d) Enlisting The Co-operation Of The Nonsense-Talker.

In a sense I have no further ideas about why someone's sincere claim to mean something should ever be rejected as mistaken. But I would like to discuss further a matter arising from the last suggestion. In (a) and (b) the nonsensicalist was presented as an aggressor attempting to show that someone else was talking nonsense but in (c) there was a change of emphasis: the nonsensicalist was trying to get the other to see for himself that he was talking nonsense. Let us take this seriously.

If we think of the nonsensicalist as trying *to refute an opponent*, then perhaps there is not much he can do to prove that his opponent is subject to an IOM. But if we think of him as trying to help someone out of philosophical perplexity - someone who acknowledges his perplexity and wants to be helped - things may seem more hopeful. Given that perplexity of any sort can be seen as a one-down position, the victim can be expected to want to overcome it and co-operate towards this end. One finds it said that he must, if arguing with him is to be of any use, *want to be cured*.[15] This *therapeutic* model of philosophical discussion often takes the form of a comparison with psycho-analysis.

It is hard to know how seriously nonsensicalists who talk this way wish to be taken. Certainly, their practice rarely accords with the theory. Wittgensteinians, such as Hacker and Baker, simply hand down refutations in precisely the way philosophers have always done. It would, for example, be grotesque to read their *Language, Sense and Nonsense* [16] as a set of psycho-philosophical case-histories in which Chomsky, Davidson, Dummett and other leading figures in the philosophy of language come to them wanting to be relieved of their perplexities. The position of Wittgenstein is less clear-cut, since he so often appears to be arguing with himself. One could perhaps see him as practising a kind of therapy - on himself.[17]

But whether or not the therapeutic version of nonsensicalism has ever been given much of a trial, it is certainly worth asking whether it is likely to prove more effective than more aggressive ones. Two points need to be made. First, traditional philosophers do not always feel perplexed. They often claim to have successfully negotiated any perplexity and reached definite conclusions. Not only are there epistemologists who would like to overcome scepticism, there are metaphysicians who are

---

15. *Insight and Illusion*, revised edition, Clarendon, 1986, p.247. Hacker would also say that, unfortunately, many of us remain *attached* to our illusions. (See *Analytical Commentary*, Vol.III, Blackwell, 1990, p.264)

16. Blackwell, 1984.

17. Perhaps he would say that this is an essential part of his method: that it is because he has made or been tempted to make philosophical errors himself that he recognises them in others. This would make a solution to the Problem of Diagnosis dependent upon a solution to the Problem of *Self-Diagnosis*.

confident they are right.[18] Second, it is necessary to ask by what standards therapeutic nonsensicalism should be judged. Even when philosophers admit to being perplexed, we still need to know whether the cures on offer are of any value. Are the nonsensicalist's claims to superior insight (understanding the victim of philosophical perplexity better than he understands himself) ever justified? The comparison of therapeutic nonsensicalism with psycho-analysis is surely dangerous: without wishing to join in the contemporary anti-Freudian chorus, I would at least point out that it is notoriously difficult to assess either the therapeutic value or the theoretical validity of psychoanalysis.

Suppose a philosopher is induced to reject certain of his earlier utterances as nonsensical: philosophers have certainly been known to change their minds about the meaningfulness of utterances. Could the means by which this was achieved have anything like cogency?[19] In a sense I am reiterating the question with which this chapter opened but with a different emphasis: assuming a change has been brought about *somehow*, how should it be assessed? Kierkegaard somewhere tells of a Catholic and a Protestant who argue about religion and end up converting each other. I forget the point of the anecdote but the moral I draw from it is that people can change their minds and change each other's minds but whether these changes are in any way justified, in particular whether a given change is more rational than one in the opposite direction, is quite another matter. Our discussion so far has not identified any way in which it could be shown that someone (oneself or another) is wrong to think he means anything by an utterance. We have discovered nothing resembling a cogent argument.

It seems to me that there is a danger that a philosopher who is perplexed will want to end his perplexity and will allow the nonsensicalist to foist on him stipulative notions of meaningfulness in order to bring this about. For example, many philosophers would

---

18. There is a strong flavour of the patronising about much therapeutic philosophy. Remember the joke, 'How many psycho-analysts does it take to change a light-bulb? - Only one, but the bulb must really want to be changed.' It applies equally to therapeutically-inclined philosophers.

19. According to Hacker (*Insight and Illusion*, p.208), 'Wittgenstein thought that the notion of proof in philosophy was obsolete ... Scepticism is not to be answered by proving that we do know what the sceptic doubts, but rather by showing that sceptical doubts make no sense ...' This 'showing' is not proving, presumably. What is it then? And does it have anything comparable to cogency?

like to see the back of scepticism. I myself, when I began philosophy, found epistemological questions extremely interesting (precisely what philosophy was about, I might have said) but since then my interest has waned, perhaps in part because so little genuine progress has been made with them. There is an obvious temptation to look for an excuse to dismiss them and the idea that they only *seem* meaningful might appear to be just what is needed. But such a dismissal would only be justified if one really did have grounds for thinking the appearance of meaningfulness illusory. Otherwise 'nonsense' would be little more than the label one put on the rubbish-bin to which one consigned these questions. One might think one is refusing to answer the sceptic because his doubts make no sense but perhaps one is really claiming they make no sense for no better reason than that one is unable to answer them.[20] Might one not do better simply to admit that one had lost interest in or patience with them?

But though there is a danger of self-deceptively dismissing difficult questions as nonsensical, perhaps not all cases are like this. Let me conclude this chapter by considering a case in which the nonsensicalist has something helpful to say but characteristically goes too far.

Suppose someone asks, 'Did time have a beginning?' Later he decides or is persuaded that he ought first to have asked, 'Does the expression "the beginning of time" have any meaning?' or, perhaps, 'How are we to give sense to the expression "the beginning of time"?'. This sort of thing does happen and it seems just the kind of insight the nonsensicalist wants to encourage. Has this philosopher realised he has given *no meaning* to certain signs in his earlier 'question'? Wittgenstein remarks that we have got 'to make the rules of the game before we play it[21] and it looks as though this is just what our over-hasty philosopher did not do.

And yet if one claims that at first he meant *nothing whatsoever* by 'Did time have a beginning?', one gets into trouble. For then it becomes difficult to say that the improved questions express what he was really getting at, wanting to say, trying to say, with the first.

---

20. A similar question might be raised about Russell's Theory of Types. Are we being asked to reject certain putative propositions about classes as nonsense *simply because* they lead to paradox? Or can some independent reason for making this move be found, so that we can say, 'No wonder paradoxes arose. We were trying to treat nonsense as sense.'?

21. Ambrose and Macdonald, op.cit., p.15.

Suppose that he had been at first inclined to deny that time had a beginning; suppose in particular that he had felt that it *could not* have had a beginning. Might he not now say, ' I used to deny that time could possibly have had beginning. I now see that I would have done better to say that I had no idea what could count as a beginning to time.'? If so, there was something insightful or at least insight-generating in his earlier question and something right about his earlier denial. Could one then go on insisting that he meant nothing by them? We are, after all, admitting that there was something he was getting at, wanting to say, trying to say, groping towards.[22]

Of course, his ruminations might have progressed differently. He might have come up with a way of giving sense to 'the beginning of time'; the fact that cosmologists find that certain terms in their equations become infinite 'before' the Big Bang may serve to give it a sense (a possibility I shall not attempt to evaluate). Again, do we want to say that this conclusion represents a sheer discontinuity, a breakthrough from utter nonsense - like 'Ab sur ah' - into sense and that his previous thoughts, in spite of their fruitfulness, were not such as to ensure that he meant anything by 'Did time have a beginning?'? It would surely be less misleading to say that he was now making it more precise than that he was giving it all the sense it ever had.

It is important to notice that in resisting the nonsensicalist urge to stigmatise the question as nonsensical one is not denying that meaning is prior to truth and that one cannot answer it until work has been done to give it a more precise sense. In Chapter Four, when discussing 'What time is it on the sun?', we encountered questions that manifested a certain naivety as to the prospects for answering them just as they stood and 'Did time have a beginning?' seems to be a case in point. It is possible therefore that the intense concern with meaning that has provoked the ire of many opponents of 'analytic' or 'linguistic' philosophy is quite proper, even if it has taken far too heavy-handed a form in nonsensicalism.

---

22. Hacker (*Analytical Commentary*, Vol.IV, Blackwell, 1996, p.241) writes:
> There cannot be a grain of truth in a nonsense or more truth in some nonsenses than in others, for nothing is said by a nonsense. Nor can there be depth, importance, or illumination to a meaningless conjunction of words, even if it possesses the *Satzklang* of a sentence of the language. What is true is that the *motivations* behind many nonsensical philosophical assertions *may* include a grain of truth or an insight that is distorted.

Surely if one makes this concession about the motivation of a philosopher, one is not well-placed to insist that he doesn't mean *anything* by what he says (though it might be true that what he says is not an acceptable sentence of the language).

Chapter Ten: What Are We To Conclude About Nonsensicalism?

I t will be evident that I do not consider the case for nonsensicalism to be particularly strong. Still, a few things can be said in its favour:

a) We found in Chapter Six that those who seem to experience revelations after taking drugs are often, once the effects have worn off, unable to understand what they themselves said or wrote under their influence. This is on the face of it evidence that they were victims of IOMs. It is possible that they have simply forgotten what they meant, but it seems somewhat high-handed to insist that this must be so. I left the question open, saying that it was cases of this sort, not anything said by philosophers, that struck me as the strongest reason for thinking that we should try to allow for the possibility of IOMs.

b) The continued intractability of philosophical problems has suggested to many that there is something wrong with them. I warned (Chapter Four) that it would invite confusion to use the word 'nonsense' as a catch-all for any kind of question that has something wrong with it. But there is no doubt that many philosophers intend their claims that philosophical problems are nonsense to be taken in the strict sense with which I have been concerned. I noted (Chapter Five) that the view that they will turn out to be nonsensical does not seem to have been more successful in providing agreed *dissolutions* of them than the old view that they are perfectly meaningful was in providing agreed *solutions*.

Let me say a little more about the intractability of philosophical problems. Perhaps the underlying difficulty, though we may be reluctant to admit it, is that we have little idea what the successful resolution of a philosophical problem would look like.[1] (I use the word 'resolution' to cover both the solution of genuine problems and the dissolution of nonsensical pseudo-problems.) Nor, and this is a closely related point, have we much idea what makes a wrong philosophical view - a mistaken answer to a philosophical problem - wrong. There are certainly philosophical doctrines that

---

1. The case for this is well presented in Colin McGinn's *Problems in Philosophy: The Limits of Inquiry*, Blackwell, 1993. I venture no opinion on his view that human beings are fated to remain forever in this predicament

are no longer widely held, and perhaps a few that no one now holds, but whether they are contingently false, self-contradictory, necessarily false in some other way than by self-contradiction, nonsensical, incapable of explaining what they are supposed to explain, or something else is not a point about which one would expect much agreement.[2]

It is true that philosophical problems have so far proved intractable but where do we go from there? Nonsensicalism tries to treat their intractability as pointing the way towards their resolution: they are intractable because they are nonsensical and we should give up trying to solve them and try to dissolve then instead. But since this approach has not proved very successful, they remain intractable. Nevertheless, so long as this situation persists it will be reasonable to take seriously the possibility that there is something wrong with philosophical problems (or some of them) and one form that this suspicion takes is nonsensicalism.

c) When we enquired how IOMs might be possible we found an answer of sorts. The suggestions reviewed in Chapter Seven were too undeveloped to be convincing, but in Chapter Eight it was suggested that an account based on Wittgenstein's view of what meaning is *not* was more promising. Since meaning something by an utterance is not, according to him, a matter of what is going on in one's mind at the time, anything that happens to be going on when there is genuinely something one means might also go on when there is not. This makes it easier to deny that a person is the final authority on whether he means anything. But everything now hangs on why one should want to overrule someone's sincere claim that there is something he means. One has created the conceptual space for doing so; could one have grounds for it? It was hard to see how one could

---

2. It is noteworthy that there is far more agreement among philosophers about the assessment of *arguments* than about the assessment of *theses*. I do not mean that all philosophers agree about the validity of a particular argument of, say, Berkeley's; but that if two philosophers agree about the *form* of the argument, they are likely to agree about its validity. By contrast they can easily agree about the interpretation of the thesis the argument is intended to establish without agreeing whether the thesis is right or wrong, and even if both think it wrong, they need not agree about what its wrongness consists in. Modern philosophers can no doubt do a pretty good demolition job on Berkeley's arguments for Idealism. But suppose we ask, 'Setting his arguments aside, could his Idealism be true; if not, why not?' I do not think there would be any agreed answer to that.

WHAT CAN WE CONCLUDE ABOUT NONSENSICALISM?

(Chapter Nine). Explaining an illusion of any sort will face two challenges: explaining the appearance and explaining why the appearance is to be dismissed as misleading. We made some progress with the first only to find ourselves stumped by the second.

Such then is, in my view, the best that can be said for the belief in IOMs. Clearly, if I am right, the confidence with which nonsensicalists have assumed their possibility is unjustified. Let me now summarise the case against nonsensicalism. Rather then simply listing difficulties and oddities, I shall try to present them so as to bring out how the underlying problems connect with each other.

When discussing criteria of meaningfulness, specifically verificationist ones, in Chapter Nine I suggested that those employing them might be employing their own stipulative notions of what meaningfulness is. Suppose this were true of nonsensicalists generally. It would explain a great deal. It was noted in Chapter Five that there are no uncontroversial examples of philosophical nonsense and that locutions normally continue to *seem* meaningful to those who have decided that they constitute philosophical nonsense. If nonsensicalists were engaged in stipulation, this would not be surprising. Nor would their tendency to slip into treating as meaningful what they want to claim is meaningless - for example, by asking what it entails. When they seem to be unblushingly treating someone as meaning something by an utterance in order to show that he means nothing by it, perhaps what they are really doing is trying to show that he means nothing, where meaning something by what one says is being conceived more narrowly than usual.[3]

Consider the following fictitious case. A philosopher claims that tautologies and contradictions are nonsense. He accuses someone of talking nonsense on the grounds that what he says is contradictory and offers a proof that it is which pays careful attention to what the accused *means* by it. (For example, he might take trouble to show that the accused is using a certain word *univocally*.) Now Wittgenstein in the *Tractatus* avoided this muddle by creating a

---

3. In Chapter Four I mentioned the possibility that the concept of meaning something by what one says is a vague one: there might be border-line cases. This of course would license stipulation, but not just any stipulations. My suggestion is that nonsensicalists are introducing quite novel conceptions of what it is for there to be something one means by what one says, not just tidying up troublesome areas of indefiniteness in the ordinary conception. Such tidying up would hardly have the revolutionary consequences of the idea that philosophical questions and answers are nonsensical.

special category for contradictions and tautologies so as to avoid characterising them as sheer nonsense: they are senseless rather than nonsensical (4.461, 4.4611). But our philosopher is less circumspect. Yet one can see why he wants to call contradictions and tautologies nonsensical: they are, in an important sense, *empty*, one part of a contradiction or tautology seeming to cancel out the other part. My suggestion is that other nonsensicalists tend to make a less obvious version of this mistake (not always *much* less obvious). They have objections, which may or may not be well-founded, to certain utterances understood in certain ways but they confusedly and confusingly express those objections by denying that the utterances have any meaning to be understood.[4] What they are really doing is denying that the utterances have any meaning, where meaning is conceived in some narrower way than was needed when formulating the objections.[5]

It will no doubt be replied that this may sometimes happen but it does not show that all nonsensicalists are confused in this way. In particular, nonsensicalists who are not verificationists may cheerfully agree that the verificationists are just insisting on their own excessively narrow conception of meaning whilst having to employ a more normal conception when testing an utterance for verifiability. But they will claim that there is more substance to their own

---

4. One could say that, just as many of the contributors to *The New Wittgenstein* (ed. Alice Crary and Rupert Read, Routledge, 2000) urge an 'austere' view of nonsense, I am insisting that we take an austere view of sense: we should not pretend that we are not according sense to something when we are. Edward Witherspoon argues that many nonsensicalists are forced into the untenable position of attributing 'quasi-meanings' to utterances they claim are nonsense. (Ibid., 'Conceptions of nonsense in Carnap and Wittgenstein', pp.339-42) My suggestion is that they are unwittingly attributing ordinary meanings to them. The difference between us is, of course, linked to the fact that he believes there are ways of diagnosing the talking of nonsense which avoid the traps into which so many nonsensicalists have fallen whereas I am sceptical about this.

5. Kingsley Amis (*The King's English*, Harper Collins, 1997, pp.131-2), while discussing what he takes to be misuses of the word 'meaningful', points out that the word did not exist in ordinary English until recently. I find this surprising - it seems altogether natural to have a word for the opposite of 'meaningless' - but examination of older dictionaries seems to bear him out. What I find more remarkable however is that not only does he treat the word as originally a technical term of philosophy, he glosses the technical usage as 'verifiable or disprovable', not as 'having meaning'. Clearly he has in mind the positivists but what is significant is that he sees them not as trying to bring out what is generally understood by 'meaning' but as simply introducing a technical term.

accusations of talking nonsense. They will claim that those they accuse do not mean anything in the ordinary sense of 'mean', not in some technical sense. But they will then face the difficulties encountered in sections (b), (c) and (d) of Chapter Nine. We were unable to find any way in which the nonsensicalist could show that the accused had given no meaning to the words he was using. Even if the accused is prepared to accept the *possibility* that he might be wrong to think he means anything by an utterance and even if his desire to escape philosophical perplexity provides him with a motive for rejecting his own utterances as nonsensical, there seems no way of *demonstrating* to him that he means nothing. The nearest we got to this was with questions manifesting a certain naivety about the possibility of answering them just as they stood and here it could not be said that someone asking them did not mean anything whatsoever. The claim to be using the ordinary concept of meaning is suspect for a further reason: it is far from clear that the distinction between meaning something by what one says and merely thinking there is something one means has any currency outside philosophy.

In Chapters Three and Four I wondered whether there could be a rigorous nonsensicalism which avoided treating nonsense and nonsense-talkers as meaning something. Clearly, if my suggestion about what nonsensicalists are really doing is correct, this is not *their* problem. In spite of any claims they may make to the contrary, they are allowing that those they attack mean something in a weak sense (the ordinary sense) in order to show that in some stronger sense they do not mean anything. Their problems are, first, clarifying their terminology so that 'meaning', 'nonsense' and related words are not simply used ambiguously and, second, justifying the introduction of this stronger notion of meaning. (Why use the word 'meaning' in this connexion at all?)

What though of the rigorous use of words like 'nonsense' that in Chapter Four we found in the writings of some nonsensicalists? Is it just a sham? It has to be said, I think, that the practice is quite out of step with the theory. On the one hand we are told that when a locution is senseless it is not its sense that is senseless and that philosophical nonsense is no more meaningful than 'Ab sur ah'. On the other it is unclear that actual diagnoses of talking nonsense ever do or ever could match this rigour.

Imagine a would-be rigorous nonsensicalist who says:

I am not going to put any interpretation whatsoever on what

the other is saying. And yet on the basis of what I am treating as mere sounds or marks on paper I will demonstrate that he means nothing by them.

Our discussion in Chapters Three and Nine of the Problem of Diagnosis shows why this seems hopeless. But suppose he says as a last resort:

I will not attribute any meaning to *his utterance* but that does not debar me from claiming to understand *him*. I can see what he is trying to do and demonstrate his failure without making it appear that he has succeeded.

Is this likely to be more than a verbal shuffle? What his stance would amount to would be:

I don't understand what he is saying but I understand his wanting to say it.

I submit that this could only be said by someone who was, consciously or unconsciously, employing a more than usually stringent conception of meaning and understanding what is meant.[6] In Chapter Nine we were unable to find any good reason for preferring to speak of people as being misled by false grammatical analogies, trying to make a move in one language-game that was only appropriate to another, etc., as against simply admitting that there was something they meant by what they said. So we are back with stipulation, disguised linguistic legislation, and have not achieved a rigorous nonsensicalism.

I believe that one of the main errors of those who have kept nonsensicalism alive for the last few decades is their failure to ponder the fate of verificationism and to ask whether it has anything to teach us about nonsensicalism in general. Although a great deal

---

6. Compare: 'I *disagree* with what you are saying but I understand your wanting to say it.' There are many contexts in which this might be said but they all require that one understand and accord meaning to what is said. I can think of only one case in which it might be fairly natural to speak of understanding someone's wanting to say a certain thing whilst denying that he means anything by it. There is a neurological condition known as 'Tourette's Syndrome' in which the patient suffers from various tics, often including a compulsion to utter obscenities. Suppose there were a condition in which one felt a compulsion to utter nonsense, i.e. locutions which one did not even claim to understand oneself. A fellow-sufferer could certainly claim to have some understanding of the compulsion without also claiming to understand the nonsense uttered. But what makes this case quite unlike that of claims to understand the 'temptation' to utter alleged philosophical nonsense is that the fellow-sufferer would not claim to understand the compulsion to utter the particular piece of nonsense in question, the choice as it were of that particular piece of nonsense.

was written about the verification principle, there was a tendency to treat it in isolation from other nonsensicalist claims and practices. In particular, the question whether it was possible to formulate a principle that would condemn as unverifiable metaphysical utterances but spare scientific laws dominated the discussion, so much so that one gets the impression that if this had been achieved, many philosophers, even today, would have no objection to passing from 'unverifiable' to 'meaningless'. I have no wish to dismiss the question as without interest: it would arise if someone suggested that we should eschew unverifiable claims *as such* without adding that their unverifiability made them meaningless. But suppose philosophers had laid the emphasis elsewhere; suppose they had more persistently pressed the question, 'Assuming for the sake of argument that one could demarcate a class of unverifiable utterances that did not bracket scientific laws with metaphysical speculation, would it be right to reject such utterances as *meaningless*?' I suggest that in such a case the general question of the tenability of nonsensicalism would not have been neglected.

To ask, 'Is this claim meaningless because it is unverifiable or unverifiable because of what it means?' is, at least implicitly, to put in question nonsensicalism as a whole in a way that asking whether metaphysical utterances are unverifiable in some more damaging sense than are the universal generalisations of science is not. For the nonsensicalist always faces the difficulty: how is he to avoid according meaning to the very utterance he wants to claim has no meaning and if he cannot, is he not allowing that it has meaning in the ordinary sense in order to deny that it has meaning in some narrower sense of his own invention? In the earlier chapters I was anxious to avoid any assumption of guilt by association, in particular the assumption that because verificationism (Malcolm's, for example) seemed indefensible, all forms of nonsensicalism must be. But now we can, I think, conclude that the problems facing verificationism ought to have made philosophers suspicious of all forms of nonsensicalism and that such suspicions would have proved well-founded.

Acknowledgements

I am grateful to Taylor & Francis Books Ltd. for permission to quote extensively from Norman Malcolm's *Dreaming*.

I would also like to thank Jim Grant and Steve Kupfer for their helpful comments and criticisms, and for many stimulating discussions on matters relating to the theme of this book.

Abbreviations

The following abbreviations are used for the works of Wittgenstein:

BB   *The Blue and Brown Books*, Blackwell, 1958

CV   *Culture and Value*, ed. G.H. von Wright and Heikki Nyman, trans. Peter Winch, Blackwell, 1980

OC   *On Certainty*, ed. G.E.M. Anscombe and G.H. von Wright, trans. Denis Paul and G.E.M. Anscombe, Blackwell, 1969

PG   *Philosophical Grammar*, ed. Rush Rhees, trans. A.J.P. Kenny, Blackwell, 1974

PI   *Philosophical Investigations*, ed. G.E.M. Anscombe and Rush Rhees, trans. G.E.M. Anscombe, Blackwell, 1953

TLP   *Tractatus Logico-Philosophicus*, trans. D.F. Pears and B.F. McGuinness, Routledge and Kegan Paul, 1961

Z   *Zettel*, ed G.E.M. Anscombe and G.M. von Wright, trans. G.E.M. Anscombe. Blackwell, 1967

INDEX